Brewing and Craft Beer

Brewing and Craft Beer

Special Issue Editor

Luís Ferreira Guido

MDPI • Basel • Beijing • Wuhan • Barcelona • Belgrade

Special Issue Editor
Luís Ferreira Guido
University of Porto
Portugal

Editorial Office
MDPI
St. Alban-Anlage 66
4052 Basel, Switzerland

This is a reprint of articles from the Special Issue published online in the open access journal *Beverages* (ISSN 2306-5710) from 2018 to 2019 (available at: https://www.mdpi.com/journal/beverages/special_issues/Craft_Beer)

For citation purposes, cite each article independently as indicated on the article page online and as indicated below:

LastName, A.A.; LastName, B.B.; LastName, C.C. Article Title. *Journal Name* **Year**, *Article Number*, Page Range.

ISBN 978-3-03921-489-1 (Pbk)
ISBN 978-3-03921-490-7 (PDF)

Contents

About the Special Issue Editor

Luís Ferreira Guido is currently an Assistant Professor in the Department of Chemistry and Biochemistry at University of Porto. He completed a M.Sc. in Biochemistry (1995) and a Ph.D. in Analytical Chemistry (2004) at University of Porto. He has published about 50 papers in SCI indexed journals and 2 book chapters. He is the co-author of one European Patent and Editorial Board Member of several journals. He has supervised 5 Ph.D. theses and 30 M.Sc. theses. His main research interests are divided into analytical chemistry (mass spectrometry, chromatography and bioanalytical techniques) and brewing chemistry (beer flavor, beer stability, antioxidants, and polyphenols).

Preface to "Brewing and Craft Beer"

Beer is a beverage with more than 8000 years of history, and the process of brewing has not changed much over the centuries. However, important technical advances have allowed us to produce beer in a more sophisticated and efficient way. The proliferation of specialty hop varieties has been behind the popularity of craft beers seen in the past few years around the world. Craft brewers interpret historic beer with unique styles. Craft beers are undergoing an unprecedented period of growth, and more than 150 beer styles are currently recognized. While many studies have suggested the beer value chain might be a vehicle for economic growth, few have estimated the economic impacts of craft beer to a geographical region. Craft Beer as a Means of Economic Development: An Economic Impact Analysis of the Michigan Value Chain is a very interesting study in this Special Issue suggesting that state governments might generate economic growth by creating a business climate that is conducive to the growth of the instate beer value chain [1].

This Special Issue, Brewing and Craft Beer, comprises nine different works by researchers from five continents (North America, South America, Europe, Africa, and Oceania). This Special Issue reflects thus a broad perspective on the most important questions that concern the researchers in different parts of the world. One such problem is the difficulty to cultivate barley on the African continent. Barley is a temperate cereal, and the African climate is unsuitable for its cultivation. The process of brewing lager beer with cereals other than barley is growing to be a common practice, especially in non-barley-producing countries. Mbeh Harry et al. demonstrate how sorghum can be well and efficiently utilized industrially in Cameroon for producing beer [2].

Hops are a very important component for the success of craft beer. The aroma and flavor of hops in strongly hopped and often in dry hopped beers are particularly responsible for the character of such beers. The work of da Costa Jardim et al. [3] advances the understanding and complexity of the sensory profile of different styles of craft beers from Southern Brazil. They report that the beer with the lowest bitterness had a higher preference among consumers, showing bitterness as a key factor that influences beer preference and leads to a decline in consumer preference. Also from Brazil, a comparative study of dry and wet milling of barley is presented by Pereira de Moura et al. [4]. Their results indicate the best milling conditions to obtain a good mashing yield in order to increase competitiveness of the microbreweries sector, as well as to improve product quality and to promote the reduction of production costs.

The main quality characteristics of beer are appearance, aroma, flavor, and mouthfeel. Computer vision is a non-destructive technique which has been applied in automated inspection and measurement. Lukinac et al. present an overview of the applications and the latest achievements of computer vision methods in determining the quality attributes of beer [5]. The use of machine learning algorithms, especially artificial neural networks (ANN), has become more popular in recent years, as they aid in increasing accuracy and reducing time and cost through analytical and sensory methods to assess the quality and acceptability of beverages. These models may be used as a cost-effective method for the fast-screening of beers during processing to assess the acceptability more efficiently, as reported by Gonzalez Viejo et al. [6]. The same team uses a machine learning modelling approach to study the effect of soundwaves on foamability properties and sensory of beers [7]. They show that the use of soundwaves is a potential treatment in brewing to improve beer quality by increasing the number of small bubbles and the foamability without disrupting yeast or modifying the aroma and flavor profile.

In the brewing process, the efficiency of fermentation and the character and quality of the final product are intimately linked. Two papers are published in this Special Issue addressing the fermentation step. Firstly, Kumar et al. showed that the use of water obtained from the soaking of corn germ resulted in a shortening of the fermentation time [8]. The addition of germ water, rich in micronutrients and soluble proteins, increased the free amino nitrogen levels and Zn concentration in the wort, enhancing its economic value. Then, last but certainly not least, Silva Ferreira et al. answer the question why craft brewers should be advised to use bottle refermentation to improve late-hopped beer stability [9]. As bottle refermentation is widely used in Belgian craft beers, the aim of their work is to assess how this practice might impact their flavor. It was with great pleasure that I carried out the coordination of this Special Issue of Beverages on the topic Brewing and Craft Beer. I firmly believe that this research field will hold a very important place in the strategy of Beverages. I am especially thankful to all authors who have generously shared their scientific knowledge and experience with others through their contribution to this Special Issue. I wish to extend my thanks to the Editorial Office of Beverages, in particular to Ms. Tina Tian.

Conflicts of Interest: The author declares no conflict of interest.

References

1. Miller, S.R.; Sirrine, J.R.; McFarland, A.; Howard, P.H.; Malone, T. Craft Beer as a Means of Economic Development: An Economic Impact Analysis of the Michigan Value Chain. *Beverages* **2019**, *5*, 35.

2. Mbeh. Harry, F.; Zangué Steve Carly, D.; Emmanuel Jong, N. Sorghum Coffee–Lactose Stout Production and Its Physico-Chemical Characterization. *Beverages* **2019**, *5*, 20.

3. da Costa Jardim, C.; de Souza, D.; Cristina Kasper Machado, I.; Massochin Nunes Pinto, L.; de Souza Ramos, R.; Garavaglia, J. Sensory Profile, Consumer Preference and Chemical Composition of Craft Beers from Brazil. *Beverages* **2018**, *4*, 106.

4. Pereira de Moura, F.; Rocha dos Santos Mathias, T. A Comparative Study of Dry and Wet Milling of Barley Malt and Its Influence on Granulometry and Wort Composition. *Beverages* **2018**, *4*, 51.

5. Lukinac, J.; Mastanjević, K.; Mastanjević, K.; Nakov, G.; Jukić, M. Computer Vision Method in Beer Quality Evaluation—A Review. *Beverages* **2019**, *5*, 38.

6. Viejo, C.G.; Torrico, D.D.; Dunshea, F.R.; Fuentes, S. Development of Artificial Neural Network Models to Assess Beer Acceptability Based on Sensory Properties Using a Robotic Pourer: A Comparative Model Approach to Achieve an Artificial Intelligence System. *Beverages* **2019**, *5*, 33.

7. Gonzalez Viejo, C.; Fuentes, S.; Torrico, D.; Lee, M.; Hu, Y.; Chakraborty, S.; Dunshea, F. The Effect of Soundwaves on Foamability Properties and Sensory of Beers with a Machine Learning Modeling Approach. *Beverages* **2018**, *4*, 53.

8. Kumar, D.; Hager, A.S.; Sun, A.; Debyser, W.; Javier Guagliano, B.; Singh, V. Improving Fermentation Rate during Use of Corn Grits in Beverage Alcohol Production. *Beverages* **2019**, *5*, 5.

9. Silva Ferreira, C.; Bodart, E.; Collin, S. Why Craft Brewers Should Be Advised to Use Bottle Refermentation to Improve Late-Hopped Beer Stability. *Beverages* **2019**, *5*, 39.

Luís Ferreira Guido
Special Issue Editor

Editorial
Brewing and Craft Beer

Luis F. Guido

Department of Chemistry and Biochemistry, Faculty of Sciences, University of Porto, 4169-007 Porto, Portugal; lfguido@fc.up.pt; Tel.: +351220402644

Received: 3 August 2019; Accepted: 12 August 2019; Published: 16 August 2019

Beer is a beverage with more than 8000 years of history, and the process of brewing has not changed much over the centuries. However, important technical advances have allowed us to produce beer in a more sophisticated and efficient way. The proliferation of specialty hop varieties has been behind the popularity of craft beers seen in the past few years around the world. Craft brewers interpret historic beer with unique styles. Craft beers are undergoing an unprecedented period of growth, and more than 150 beer styles are currently recognized. While many studies have suggested the beer value chain might be a vehicle for economic growth, few have estimated the economic impacts of craft beer to a geographical region. *Craft Beer as a Means of Economic Development: An Economic Impact Analysis of the Michigan Value Chain* is a very interesting study in this Special Issue suggesting that state governments might generate economic growth by creating a business climate that is conducive to the growth of the instate beer value chain [1].

This Special Issue, *Brewing and Craft Beer*, comprises nine different works by researchers from five continents (North America, South America, Europe, Africa, and Oceania). This Special Issue reflects thus a broad perspective on the most important questions that concern the researchers in different parts of the world. One such problem is the difficulty to cultivate barley on the African continent. Barley is a temperate cereal, and the African climate is unsuitable for its cultivation. The process of brewing lager beer with cereals other than barley is growing to be a common practice, especially in non-barley-producing countries. Mbeh Harry et al. demonstrate how sorghum can be well and efficiently utilized industrially in Cameroon for producing beer [2].

Hops are a very important component for the success of craft beer. The aroma and flavor of hops in strongly hopped and often in dry hopped beers are particularly responsible for the character of such beers. The work of da Costa Jardim et al. [3] advances the understanding and complexity of the sensory profile of different styles of craft beers from Southern Brazil. They report that the beer with the lowest bitterness had a higher preference among consumers, showing bitterness as a key factor that influences beer preference and leads to a decline in consumer preference. Also from Brazil, a comparative study of dry and wet milling of barley is presented by Pereira de Moura et al. [4]. Their results indicate the best milling conditions to obtain a good mashing yield in order to increase competitiveness of the microbreweries sector, as well as to improve product quality and to promote the reduction of production costs.

The main quality characteristics of beer are appearance, aroma, flavor, and mouthfeel. Computer vision is a non-destructive technique which has been applied in automated inspection and measurement. Lukinac et al. present an overview of the applications and the latest achievements of computer vision methods in determining the quality attributes of beer [5]. The use of machine learning algorithms, especially artificial neural networks (ANN), has become more popular in recent years, as they aid in increasing accuracy and reducing time and cost through analytical and sensory methods to assess the quality and acceptability of beverages. These models may be used as a cost-effective method for the fast-screening of beers during processing to assess the acceptability more efficiently, as reported by Gonzalez Viejo et al. [6]. The same team uses a machine learning modelling approach to study the effect of soundwaves on foamability properties and sensory of beers [7]. They show that the use of soundwaves is a potential treatment in brewing to improve beer quality by increasing the number of small bubbles and the foamability without disrupting yeast or modifying the aroma and flavor profile.

In the brewing process, the efficiency of fermentation and the character and quality of the final product are intimately linked. Two papers are published in this Special Issue addressing the fermentation step. Firstly, Kumar et al. showed that the use of water obtained from the soaking of corn germ resulted in a shortening of the fermentation time [8]. The addition of germ water, rich in micronutrients and soluble proteins, increased the free amino nitrogen levels and Zn concentration in the wort, enhancing its economic value. Then, last but certainly not least, Silva Ferreira et al. answer the question why craft brewers should be advised to use bottle refermentation to improve late-hopped beer stability [9]. As bottle refermentation is widely used in Belgian craft beers, the aim of their work is to assess how this practice might impact their flavor.

It was with great pleasure that I carried out the coordination of this Special Issue of *Beverages* on the topic *Brewing and Craft Beer*. I firmly believe that this research field will hold a very important place in the strategy of *Beverages*. I am especially thankful to all authors who have generously shared their scientific knowledge and experience with others through their contribution to this Special Issue. I wish to extend my thanks to the Editorial Office of *Beverages*, in particular to Ms. Tina Tian.

Conflicts of Interest: The author declares no conflict of interest.

References

1. Miller, S.R.; Sirrine, J.R.; McFarland, A.; Howard, P.H.; Malone, T. Craft Beer as a Means of Economic Development: An Economic Impact Analysis of the Michigan Value Chain. *Beverages* **2019**, *5*, 35. [CrossRef]
2. Zangué Steve Carly, D.; Emmanuel Jong, N. Sorghum Coffee–Lactose Stout Production and Its Physico-Chemical Characterization. *Beverages* **2019**, *5*, 20. [CrossRef]
3. da Costa Jardim, C.; de Souza, D.; Cristina Kasper Machado, I.; Massochin Nunes Pinto, L.; de Souza Ramos, R.; Garavaglia, J. Sensory Profile, Consumer Preference and Chemical Composition of Craft Beers from Brazil. *Beverages* **2018**, *4*, 106. [CrossRef]
4. Pereira de Moura, F.; Rocha dos Santos Mathias, T. A Comparative Study of Dry and Wet Milling of Barley Malt and Its Influence on Granulometry and Wort Composition. *Beverages* **2018**, *4*, 51. [CrossRef]
5. Lukinac, J.; Mastanjević, K.; Mastanjević, K.; Nakov, G.; Jukić, M. Computer Vision Method in Beer Quality Evaluation—A Review. *Beverages* **2019**, *5*, 38. [CrossRef]
6. Viejo, C.G.; Torrico, D.D.; Dunshea, F.R.; Fuentes, S. Development of Artificial Neural Network Models to Assess Beer Acceptability Based on Sensory Properties Using a Robotic Pourer: A Comparative Model Approach to Achieve an Artificial Intelligence System. *Beverages* **2019**, *5*, 33. [CrossRef]
7. Gonzalez Viejo, C.; Fuentes, S.; Torrico, D.; Lee, M.; Hu, Y.; Chakraborty, S.; Dunshea, F. The Effect of Soundwaves on Foamability Properties and Sensory of Beers with a Machine Learning Modeling Approach. *Beverages* **2018**, *4*, 53. [CrossRef]
8. Kumar, D.; Hager, A.S.; Sun, A.; Debyser, W.; Javier Guagliano, B.; Singh, V. Improving Fermentation Rate during Use of Corn Grits in Beverage Alcohol Production. *Beverages* **2019**, *5*, 5. [CrossRef]
9. Silva Ferreira, C.; Bodart, E.; Collin, S. Why Craft Brewers Should Be Advised to Use Bottle Refermentation to Improve Late-Hopped Beer Stability. *Beverages* **2019**, *5*, 39. [CrossRef]

Article

Craft Beer as a Means of Economic Development: An Economic Impact Analysis of the Michigan Value Chain

Steven R. Miller [1], J. Robert Sirrine [2], Ashley McFarland [3], Philip H. Howard [4] and Trey Malone [1,*]

[1] Department of Agricultural, Food and Resource Economics, Michigan State University, East Lansing, MI 48824, USA; mill1707@msu.edu
[2] Community, Food & Environment Institute, Michigan State University Extension, East Lansing, MI 48824, USA; sirrine@msu.edu
[3] Business and Industry Liaison, Minnesota Sea Grant, Duluth, MN 55812, USA; ashleymc@d.umn.edu
[4] Department of Community Sustainability, Michigan State University, East Lansing, MI 48824, USA; howardp@msu.edu
* Correspondence: tmalone@msu.edu

Received: 26 February 2019; Accepted: 15 April 2019; Published: 2 May 2019

Abstract: While many studies have suggested the beer value chain might be a vehicle for economic growth, few have estimated the economic impacts of craft beer to a geographic region. As such, this study uses modified input/output analysis to identify the economic contributions of instate beer production to the Michigan economy. We find that the beer value chain generated nearly $500 million in Gross State Product in 2016, contributing nearly $1 billion as well as 9738 jobs in total aggregate economic contributions. The results suggest that state governments might generate economic growth by creating a business climate that is conducive to the growth of the instate beer value chain.

Keywords: craft beer; local value chain; economic contribution analysis

1. Introduction

Where past economic development efforts focused on attracting new industries to create jobs, economists now recognize the value of import substitution as a possible method for developing regional economies [1]. The "Buy Local" movement is one key embodiment of import substitution, as increasing purchases within one's own region reduces leakages within an economic value chain, while simultaneously increasing producer surplus [2]. Corresponding with the entry of a younger generation of consumers, the notion of "consumer-driven growth" has also entered the economic development lexicon [3,4]. Consumer-driven growth is increasingly important, as many consumers desire a proliferation of high-status options [5]. The economic theories of import substitution and consumer-driven growth converge perfectly in the craft beer value chain. Many consumers are willing to pay higher prices for more unique craft beer varieties, signaling that state governments might create a business climate that is conducive to economic growth via the beer value chain [6]. This is especially important as governments have sweeping power to regulate the production, importation and sale of beer [7].

It follows that policymakers would benefit from understanding the economic contributions of the craft beer value chain. Sometimes state legislation can impede the formation of local and regional breweries, thereby impeding its entrepreneurial activity [6,8]. In response to increased consumer interests in craft beer, many states have recently provided further opportunities for the growth of local and regional craft beer by restructuring their beer laws [9]. A thriving local craft beer industry can contribute to local and regional economic development [10]. It is in this light that we assess the economic contributions of Michigan's growing beer sector.

This article provides an estimate of the economic impacts associated with a regionally exclusive beer value chain, largely made up by craft beer. The Michigan beer value chain makes an ideal case study for our analysis as it ranks fourth in the nation in terms of craft breweries and 11th overall in terms of craft beer production [11]. What started as three microbreweries in the state in 1993 has now grown to over 330, with expectations for future growth. Michigan's long history of beer production includes a similar history in the production of key inputs, including malting barley and hops [12]. In response to brewery demand, Michigan agricultural producers have been rapidly increasing the amount of farmed malting barley and hop acres, where Michigan is developing a regional reputation for quality beer inputs. Specifically, hops grown in Michigan have clear attributes that differentiate varieties from those grown in other hop producing states, namely Washington, Idaho, and Oregon. Michigan's brewers are interested in sourcing barley and hops from local sources and generating a genuinely Michigan made product [12]. In addition, a cadre of specialty malt producers has developed to help meet in-state demand. In addition to yeast harvesting and cultivation, wheat and rye are also grown in Michigan, which provides additional potential for a nearly complete local agricultural supply chain for beer production.

We contribute to the literature in two ways. First, we fill a gap in the economic development literature by generating an estimate of the economic contributions of a local food system that might be accessible to all states. By focusing on the beer value chain, we contribute to the literature in a second way. The craft beer market has been identified as a vehicle for economic development, but studies on the value chain economic contributions of state-level beer production are lacking. The remainder of this article is organized as follows: First, we provide a brief background of the beer value chain. Second, we describe the conceptual framework of input-output analysis, which utilizes a sequence of interlinking production functions to estimate the total (direct, indirect and induced) economic impacts of the beer value chain in Michigan. Following the conceptual framework, we explain our application of input-output analysis to the Michigan beer value chain. We then describe the results of our model, which suggest that the beer value chain contributed nearly $500 million to the Gross State Product in 2016. The final section concludes with a discussion of the implications and limitations of the current study.

1.1. Background

The industrial organization of the U.S. beer industry has an incredible history [13]. In 1873, 4131 breweries operated in the United States. Industry consolidation reduced the number of breweries to 100 in 1978, operated by only 50 firms, as brewers tended toward flavors with the broadest consumer appeal [14]. This developed a marketplace where brand loyalty is fierce even though most consumers cannot distinguish between leading brands [15]. The beer market changed significantly in 1979 when changes in federal laws freed states to allow for homebrewing [13]. Over the next few decades, many of those hobbyists set out to commercialize their homebrews, planting the seeds of a burgeoning craft beer market [16].

The Brewers Association defines craft breweries as independently owned and producing less than 6 million barrels a year [17]. They further define Microbreweries as producing less than 15,000 barrels a year, with at least 75 percent of sales off-premise—though Michigan law defines microbreweries as those producing less than 60,000 barrels per year. Brewpubs sell more than 25 percent or more of its beer on site. Today, over 7000 breweries operate in the U.S., but market share is still concentrated with a few national firms. There are about 15.5 breweries per million residents today compared to 96.1 per million in 1873. The six largest firms make up about 84.5 percent of the total market of beer sales, led by Anheuser-Busch InBev, with 40.2 percent of the market [18]. Craft beer consumers largely fall into one of five categories: Traditional beer drinkers, mavens, locavores, premium beer drinkers and uninformed beer drinkers [19].

Brewers compete in one of three markets: (1) Local, (2) regional or (3) national. As most brewers start out small and with a local footprint, they start out in a local market. Upon developing a following in their local market and depending on the span of this market, some expand regionally and sell outside of the state. Reaching this level requires a level of distribution resources that are usually reserved for the top tier national brands. However, the obstacles to reaching new markets are getting smaller through information technologies and wholesalers that can take a product to a national market with little up-front costs to the brewer. Conversely, a brand can reach a national audience through acquisition by a national supplier. Acquisitions have become common in the craft beverage industry, although beer labeling is not always clear about the transition of ownership. This is due in large part to the value beer drinkers place on independently owned breweries [20]. For a more thorough discussion of mergers and acquisitions in the U.S. beer market, see Howard [21].

1.2. Conceptual Framework

Input-output (I-O) models have become staple economic impact models for regional analysis and are often used in generating economic impact estimates of publicly-funded projects [22,23]. Because I-O models trace transactions across industries and institutions, they are instrumental in understanding how changes in one sector of a region's economy can affect all other economic sectors within the region [24]. They have been especially popular in estimating the economic impacts of local food systems [25–30]. Their applicability to a myriad of questions has resulted in I-O models being the most applied economic modeling approach used in economic analysis [31].

Input-output models generalize linear transactions in a set of multipliers that capture the full extent of transactions associated with any changes in the level of production in an industry. Mathematically, the total effect of this change can be specified as:

$$\text{Total Effect} = \text{Direct Effect} + \text{Indirect Effect} + \text{Induced Effect} \tag{1}$$

The I-O model takes changes in demand (the "direct" effect) and relates them to overall economic impact (the "total" effect) through a set of mathematical equations. The indirect effect is the value of secondary inter-industry transactions in response to direct effects. The "induced" effect is the value of transactions resulting from changes in income in response to direct effects. Because the relationships are linear, the direct, indirect and induced effects can be specified as multiples of the direct effect, so Equation (1) can be restated as:

$$\text{Total Effect} = (1 + k_1 + k_2) \cdot \text{Direct Effect} \tag{2}$$

where k_1 and k_2 are greater than or equal to zero and represent the multiplicative response of indirect and induced transactions, respectively. For simplicity, we restate Equation (2) as:

$$\text{Total Effect} = k \cdot \text{Direct Effect} \tag{3}$$

where $k = (1 + k_1 + k_2)$. Equation (2) suggests that the economy-wide impact, the total effect, is some multiple of the direct effect, where the multiplier takes a positive value that is equal to or greater than one. Consider the minimum value the multiplier can take: One. This value reflects the intuitive result that if the economy's output of agricultural products expands by \$1 million dollars, for example, the economy will expand at least by \$1 million dollars. If the indirect and induced effects are not equal to zero (i.e., not imported), this \$1 million increase in output will spur other industries to expand their output of goods and services and will generate household income that will be applied to the purchase of goods and services in the economy, generating a total economic impact greater than the initial \$1 million expansion. It follows that the economic multiplier is specified as a ratio of the total to direct effects. Rearranging Equation (3) provides:

$$k = (\text{Total Effect})/(\text{Direct Effect}) \tag{4}$$

where the multiplier, k, encompasses all the direct, indirect and induced effects for a given industry and denotes the impact of a change in direct effects on the total economic system. Each industry in a region is characterized by its own value of multiplier k. Industries with expansive localized production chains will tend to have higher multipliers than industries that rely on suppliers outside of the modeling region. When there is adequate supply within the region, the region has more potential to retain the total effects of the industry. However, when producers depend on supplies outside the region, leakage occurs, and part of the total effect is lost.

All versions of I-O models require several common restrictive assumptions. First, the model imposes constant returns to scale, such that a doubling of output requires a doubling of all inputs. Second, technology is fixed with no substitution. Combined, these two assumptions mean an increase in industry output requires an equal and proportionate increase in all inputs. Additionally, supply is assumed to be perfectly elastic such that there are no supply constraints. This final assumption also asserts that all prices are fixed, such that an increase in demand for any commodity will not result in a price change for that industry. Input-output models have been criticized on the grounds that some of these assumptions are overly restrictive and that the magnitude of the bias generated by these assumptions is greater when the industry direct effects are larger, relative the overall size of the industry [32]. Despite this criticism, I-O models have become a standard by which economic impact assessment is generated.

2. Methods

We used a value chain assessment to estimate the economic contribution of the beer value chain in Michigan. A contribution analysis is similar to an economic impact assessment [33], but asserting an economic impact to an industry's presence posits some challenges. In an economic impact assessment, the generally accepted counterfactual state is that not all direct expenditures will exist in the absence of the measured industry or business. This assumption may be applicable for assessing the economic impact of a new factory that exports to other regions but is overly restrictive for assessing the economics of a visceral industry. For example, when measuring the economic impact of a new automobile panel stamping plant, it may make sense to assume that the transactions arising from that plant would not exist in the local economy in the absence of the plant. However, when measuring the economic impact of a dispersed industry like beer production, it is very likely that many of the transactions arising from the value chain would exist in the absence of beer production. Consumers will still buy beer, but it would be imported to the region. Farmers would still produce cash crops, but it would likely not be barley for malting. Rather than asserting that the value of all associated transactions would cease to exist, as in an economic impact assessment, an economic contribution study is silent on the counterfactual state of the economy in the absence of the measured industry. The associated contribution estimates suggest the extent to which the industry contributes to the overall size of the economy.

We used the IMPLAN Pro 3.1 economic input-output modeling software with the Michigan transactions data in this assessment. The IMPLAN Pro model is widely used in economic impact and industry contribution simulations and has become a standard resource for regional economists [27]. A Michigan-calibrated IMPLAN economic impact assessment model was used to estimate the contributions that would arise through secondary transactions. Of the commercial input-output models used in economic simulations, IMPLAN is one of the most referenced in the literature. Transactions are adopted from the Bureau of Economic Analysis's benchmark input-output tables, meaning that it traces transactions across industries, households and government units and uses these measures to estimate secondary transactions arising from a given set of industry transactions [34]. In this assessment, industry transactions include the set of estimated transactions along the value chain for producing Michigan beer. Secondary transactions are those transactions not directly made by industries making up the Michigan supply chain, but rather are generated for the provisions of intermediate goods and services inputs into the supply chain as indirect effects, and expenditures from earnings by workers, proprietors and government (through fees and taxes) as induced effects. The total contributions, estimated as the sum of direct and secondary effects, generally exceed the initial infusion of economic activity, generating what is commonly referred to as a multiplier effect. That is, once accounting for secondary transactions, the total contribution is larger than the direct value of economic activities along the value chain, as predicated by economic theory [35].

The IMPLAN model is driven by the dollar value of transactions, and therefore, the resulting contributions were measured in the value of total transactions. However, estimates of employment, labor income and contributions to gross state product were made via fixed ratios to industry output. The fixed ratios are industry averages. While IMPLAN provides a high degree of industry granularity, with 528 distinct industries, barley and hops production is included in industry aggregates of grain farming and fruit farming, respectively. Additionally, brewpubs are represented in the full-service restaurants industry. Only breweries are uniquely categorized within the 528 distinct industry categories. Hence, industry aggregates may not be suitably representative of the actual transactions associated with Michigan brewing. Consequently, we modified the industry purchases, as described by Miller [12], to better represent the industries under this study.

The direct economic contribution of Michigan's brewing industry is measured through brewing activities, including the transportation costs of the final output. However, in a separate measure, we included the economic contribution beyond brewing to include retail and food accommodation sales of Michigan-brewed beer as a separate measure, netting out the production of the beer sales in our final estimates. In this, we recognize two downstream channels from brewing by which economic value is generated.

We assumed brewery sales follow two channels to consumption: Onsite consumption, which includes own- and third-party sales, or through off-premise retail sales. For on-premise consumption, own sales command higher seller margins per unit of sale since sales are not required to pass through the three-tier trade system [36], entailing a licensed wholesale distributor. All off-premise consumption is channeled through the three-tier trade system, entailing the wholesale and retail trade sectors. Only margins earned at the point of sale are captured, and those through food and drinking places account only for beer sales components. Beer that is brewed and distributed in Michigan generates value all the way to final sale for consumption, while beer that is brewed in Michigan and shipped out of state stops contributing value for the state at the state borders. The structural relationship of the modeling frame is represented in Figure 1, where exports along the value chain are implied along the value chain, as that share that does traverse downstream along the instate value chain.

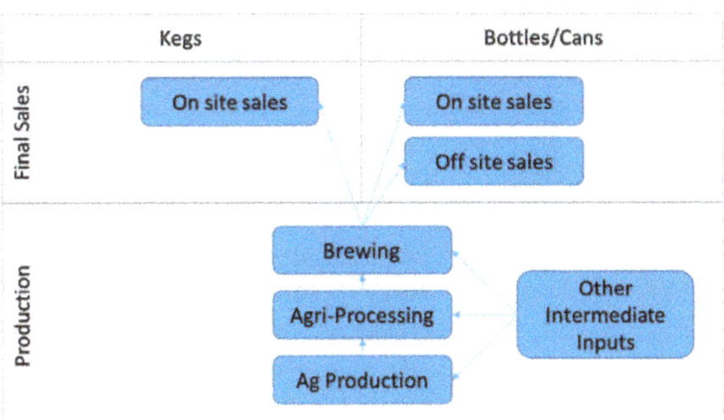

Figure 1. Michigan beer value chain.

2.1. Calculating Direct Contributions

Because all but one of Michigan's breweries were classified as craft breweries in 2016, and because the Brewers Association (BA) tracks commercial craft beer production, estimates of the production of craft beer by volume were provided by the BA statistics to be 846,029 barrels [11]. The volume of production from the one non-craft beer brewery was added to get the total volume of beer produced in-state. This beer may be packaged in barrel form for draft sales or may be packaged in bottles and cans. Using national statistics, about 61 percent of beer purchases are in the form of kegs, while the remaining 39 percent is sold by the bottle or can [37]. Additionally, the market research firm Mintel provides an estimate of where consumers purchase beer based on household surveys, showing that 79 percent of beer is consumed away from home [18]. Applying estimated sales values to volume allowed us to isolate estimates of the economic contributions of on-premise and off-premise sales channels. Assuming equitable combinations of inputs, except for beer packaging, in the brewing process and all associated agricultural inputs, the remainder of the up-stream value chain can be estimated in the aggregate based on the volume of in-state production.

Through discussions with growers and maltsters, it was determined that malting barley production largely follows standard winter and spring grain production processes, by which the IMPLAN software is readily sufficient for estimating. Similarly, as craft beer production differs from that of national brands, which are largely represented in the IMPLAN transactions table, the standard IMPLAN production function for beer brewing was modified to account for added barley and hops. More specifically, while barley use per barrel of beer is typically higher for craft beer over conventional beers [12], much of this represents a substitution of rice for barley. However, rice and barley are grouped into a single category in IMPLAN under the category of grain farming. Since barley is produced in-state while rice is not, the substitution of barley for rice represents a reduction of the intermediate imports of rice, which is not grown in Michigan, with more barley, which is grown instate. Additionally, separate accounting for wheat beers is not necessary in the IMPLAN software, as wheat and barley represent similar inputs by volume [12] under IMPLAN's grain farming sector. As both malting wheat and barley are grown in Michigan, substituting barley with wheat does not represent a shift in import shares. Finally, hop use is also heavier in craft beer, but as hops make up a minute input by volume, the overall change of hop shares in the baseline production function is minute. Hops are vegetables and therefore reflect added vegetable inputs into the production process, which falls into the IMPLAN category "vegetables and melon farming".

Under this approach, estimates exclude the contribution of imported beers into the state, where imports include domestic and foreign beers produced outside of but imported into Michigan. Hence, the estimated contributions reflect that of the beer production value chain in Michigan, but not the entirety of the economic contribution of the beer industry as a whole. One can argue that food sales through brewpubs should be included in economic contribution estimates, as food sales complement beer sales from these venues. Rather than asserting that food sold through brewpubs contributes to Michigan's beer value chain, we opted to separate the two and focus only on contributions directly arising from the production and sale of Michigan-produced beer. To be sure, on-premise sales can accrue through food and drinking establishments, but may also entail hotels, casinos, sporting and community event venues, and other recreation outlets, outside of food and drinking establishments where beer is typically sold. However, we modelled on-premise sales as if taking place in food and drinking establishments, as the underlying production functions likely better represent the actual food and drink value chains of these alternative venues. This is also a simplifying assumption, freeing us from sorting out the values of different on-premise venues that beer is sold through.

As IMPLAN is driven on sales values, not volume of production, we must convert the volume of production into actual sales values. To do this, we broke out volume into three categories: (1) On-premise drafts, (2) on-premise bottled/canned and (3) off-premise bottled/canned. Each were assigned an average price per volume, as described below. Because the IMPLAN production function for brewing beer is dominated by larger, national brand processors, the production functions were modified based on industry interviews discussed by Miller [38]. These modifications emphasize deeper ties to local supply chains of barley and hops and deemphasize rice imports through modified regional purchase coefficients. Similarly, the trade sector retail and wholesale shares of local supply from breweries is increased, as well as that for food sector inputs.

2.2. Estimating Final Sales Direct Effects

Estimates of direct effects start at the brewery level, where estimates exist for the total volume of beer produced in the state. Production direct effects are dollar-values based on the volume of beer produced in Michigan. Total volume of brewery sales comes from the Brewers Association's 2016 estimate for Michigan [11]. However, since the Brewers Association only tracks craft beer sales, and because Michigan's Founders Brewery does not fall into the Brewers' Association definition of craft beer, estimates for the Founders Brewery production must be added to the total in-state brewery output. Estimates for Founders Brewery production in 2016 were based on direct contact with the brewery [39]. Accordingly, the in-state production of beer totaled 846,688 barrels.

Generating sale values requires breaking out production between keg sales and bottle/can sales, as the average selling prices of beer depends on the packaging. We assume producer prices of $110.00 per keg (1/2 barrel) and $4.62 per six-pack of bottled or canned beer. With 61 percent of the beer produced going to kegs, keg sales total $113,626,000. Alternatively, estimated brewery bottle/can sales totaled $83,906,000. Taken together, the value of beer sales totaled $197,532,000.

Estimating the downstream contributions of wholesalers, retailers and food services requires the application of estimates of margins earned. To simplify the estimates, we assumed that all keg sales go to on-premise consumption venues, represented by IMPLAN's full-service restaurants sector. Through industry interviews, Miller [12,38] provided estimates of sales margins, which have been reproduced in Table 1. In Table 1, brewer's prices represent the sale price brewers earn per keg or 6-pack, sold through the three-tier system of a wholesaler distributor. Wholesale margins are the added value attributed to wholesalers and the entailed shipping costs. Retail margins are only applied to off-premise sales by assumption and represent that which is earned by the retailer, while food service margins, which tend to be about four times the wholesale price, is that which is earned by food service providers. By assumption, the brewer operating its own brewpub can earn food service prices. Summing the column provides the estimated consumer prices. While not shown, the estimated sales prices were $4.79 per pint of draught beer, $1.75 per bottle/can for off-premise consumption and $4.83 per bottle/can for on-premise consumption.

Table 1. Beer prices at points of sale.

	On-Premise Consumption		Off-Premise Consumption
	Keg	**6-Pack**	**6-Pack**
Brewers' Price	$110.00	$4.62	$4.62
Wholesale/Shipping Margins	$38.50	$2.63	$2.63
Retail Margin			$3.26
Food Service Margin	$445.50	$21.74	
Consumer Price	$594.00	$28.98	$10.50

Next, the total volume sold through the value chain must be estimated based on the estimated volume of Michigan's production. First, we estimated the total volume and value of sales based on production. From which, we applied estimates of the share that remained in Michigan for final consumption. The share of beer that is sold out of state does not contribute to downstream economic impacts but increases brewery production contributions. Table 2 shows the estimated volume and sales revenues by sales channel, based on the volume of in-state beer production. As discussed above, 61 percent of volume goes to kegs, while 39 percent goes to bottles and cans. A simplifying assumption asserts that all keg production is sold through on-premise drinking establishments. Of the bottle/can-packaged volume, Mintel estimates indicate that consumers purchase 48 percent of beer for on-premise consumption and 52 percent for off-premise. Applying the prices above determines total expenditures, where brewers' earnings through keg sales topped out at $113,626,000, while that for bottled/canned production was $83,906,000. These estimates are based on the volume of production at the brewery and therefore do not account for exports.

Table 2. Revenues at points of sale with no exports.

	On-Premise Consumption		Off-Premise Consumption
	Keg	6-Pack	6-Pack
Kegs (Units)	544,370		
6-packs (Units)		4,594,123	4,976,967
Wholesale ($)	$20,958,234	$12,059,574	$13,064,539
Consumer Sales ($)	$242,516,710	$99,853,273	$16,200,028

3. Results

The final estimates of the economic contribution of beer sales net out in-state production exported out of Michigan. The most current IMPLAN data for Michigan at the time of this study was 2013. We used IMPLAN regional purchase coefficients for Michigan [34] to estimate the share of beer produced in Michigan that stays in Michigan. The IMPLAN software estimates regional trade flows using a constrained gravity model that combines trade flow data with economic intensity and is believed to be a superior approach for measuring trade flows across regions by commodity. The reference data for Michigan suggests that about 52.7 percent of Michigan's beer production remains in-state. While the underlying trade flow estimates were made three years prior to this 2016 data assessment, we have no impression that this share is either increasing or decreasing. While consumers of the growing craft beer segment express a preference for locally-sourced beers [40], Michigan producers have indicated an increase in exporting activities [12]. Hence, the final estimates of economic contributions used 52.7 percent as the basis of volume that remains in state. This rate was applied to all channels of sales, which is consistent with the input-output literature for applying regional purchase coefficients [41].

3.1. Economic Contribution Estimates

Using the assumptions described above, we estimated the economic contribution of Michigan's beer production value chain in three parts: (1) Beer brewing, (2) retail for off-premise consumption, and (3) sales for on-premise through food and beverage establishments. The findings for each contribution area are presented below. This is followed by the aggregate gross contribution estimates across the value chain.

3.2. Brewing Activity

Estimates of the contributions of brewing activities entail all the upstream transactions associated with brewing, including the purchase of inputs and services in commercial breweries. Estimates include those contributions from in-state consumption, as well as for exportation out of the state. The estimates also aggregate brewing activities for both on-premise and off-premise sales. Accordingly, we estimate the value of Michigan beer production to be $197.5 million in 2016. Once accounting for the indirect and induced effects, this and the associated transactions are expected to give rise to $314.6 million in economy-wide transactions, as shown in Table 3. These transactions are expected to give rise to some 877 new jobs per year, where 218 are directly employed in the upstream brewing value chain. These jobs support some $44.3 million in labor income in the state and contribute about $106.9 million in Gross State Product.

Table 3. Economic contribution of brewing activities.

Impact Type	Employment	Labor Income	Gross State Product	Output
Direct Effect	218	$9,931,000	$50,298,000	$197,532,000
Indirect Effect	438	$24,893,000	$39,468,000	$87,128,000
Induced Effect	221	$9,510,000	$17,094,000	$29,958,000
Total Effect	877	$44,334,000	$106,861,000	$314,618,000

3.3. Off-Premise Consumption Sales

The impacts of sales for off-premise consumption include only those sales that remain in Michigan. The estimates include both retail and wholesale sales, where only margins (markup earned by retailers and wholesalers) are used as a basis of impacts. Because of the nature of the model, the purchases of beer from brewers and wholesalers cannot be easily subtracted from the estimates. Hence, the estimated secondary effects include the expected purchases of beer through conventional channels and may not represent the true expected extent to which final transactions take place. A correction is provided in the aggregate contribution estimates below.

Retail sales margins generate a more moderate level of contribution to the Michigan economy. About $29.3 million in wholesale and retail margins is generated by Michigan-produced beer, giving rise to some $53.1 million in transactions throughout the economy (Table 4). This facilitates about 458 Michigan jobs, where about 294 are directly related to the retail and wholesale efforts of beer distribution. These jobs are expected to bring in about $19.5 million in annual labor income and contribute about $33.6 million to the Gross State Product.

Table 4. Economic contribution of retail and associated wholesale activities.

Impact Type	Employment	Labor Income	Gross State Product	Output
Direct Effect	294	$11,829,000	$19,896,000	$29,265,000
Indirect Effect	67	$3,470,000	$6,188,000	$10,704,000
Induced Effect	97	$4,177,000	$7,508,000	$13,158,000
Total Effect	458	$19,476,000	$33,591,000	$53,127,000

3.4. On-Premise Consumption Sales

The final contribution estimates are for sales for on-premise consumption and associated wholesale activities. The value of wholesaling activities differs from that of the prior section only with regards to volume. That is, we do not differentiate wholesale per-unit values between selling to retail and selling to food and drink establishments. However, the per-unit consumer prices at food and drink establishments are much higher than at retail establishments. These higher markups are attributed to the mix of services and attributes the establishment provides, whether it be the mix of food, big-screen TVs, ambiance or sporting facilities. Hence, the values earned are not necessarily driven by the value of the beer consumed, but rather by the mix of products and ambiance afforded by the venue.

Accordingly, we estimate that the contribution of Michigan-brewed beer to food and drink establishments through margins earned and associated wholesaling is $342.4 million (Table 5). This drives additional transactions up to $625.7 million annually. Being more labor intensive, direct employment (including venue and wholesaling) is expected to top out just short of 6511 jobs. Through the multiplier effect, about 8403 jobs are supported with expected total labor income of $229.9 million, contributing about $329.1 million to the annual Gross State Product.

Table 5. Economic contribution of on-premise consumption sales and associated wholesale activities.

Impact Type	Employment	Labor Income	Gross State Product	Output
Direct Effect	6511	$140,820,000	$170,031,000	$342,370,000
Indirect Effect	749	$39,719,000	$70,389,000	$128,032,000
Induced Effect	1144	$49,312,000	$88,638,000	$155,345,000
Total Effect	8403	$229,851,000	$329,058,000	$625,747,000

3.5. Estimated Gross Contributions

The final task is to combine the contribution estimates into a single set of contribution estimates. The wholesale and retail sales contributions estimated above reflect the share of Michigan-produced beer that remains in state, and only account for the margins earned at each leg. Hence, adding the retail and wholesale margin contributions to that of brewing contributions avoids double counting. When combining the direct effects of sales, we assert that Michigan consumers spent about $569.17 million on Michigan-produced beer. According to the statistics reporting portal Statistica [42], per-capita expenditures on beer average about $229.40 nationally. Given Michigan's population of 9.951 million in 2016 [43], our estimates suggest that Michigan's per capita expenditure on Michigan-produced beer was $57.19, or about 24.0 percent of the national average total expenditure on beer. A Mintel survey of beer drinkers found that 19 percent of beer drinkers associate with purchasing craft beer [44]. Our findings' close proximity suggests an appropriate estimate, while the fact that our findings suggest a higher proportion of sales to craft can be attributed to the higher price points of locally-sourced beers.

The resulting gross contribution estimates are shown in Table 6. In this, the expected contribution of Michigan's craft beer value chain, including the Founders Brewery, totals $993.5 million, where the direct value chain transactions amount to about $569.2 million of that value. In total, just over 9700 jobs can be linked back to Michigan-produced beer, where about 7023 are directly tied to brewing, moving, selling and serving beer. The largest bulk of this is in the food and drink service industry. Total labor income is expected to be about $293.7 million, making up average annual job earnings of around $30,156. Finally, our estimates suggest that Michigan's craft beer sector contributes some $469.5 million to the annual Gross State Product.

Table 6. Aggregate economic contribution of Michigan's brewery value chain.

Impact Type	Employment	Labor Income	Gross State Product	Output
Direct Effect	7023	$162,580,000	$240,225,000	$569,166,000
Indirect Effect	1254	$68,082,000	$116,046,000	$225,864,000
Induced Effect	1461	$62,999,000	$113,240,000	$198,462,000
Total Effect	9738	$293,661,000	$469,511,000	$993,492,000

4. Conclusions

The U.S. beer industry is undergoing a significant, consumer-driven restructuring as portions of the national alcohol landscape are shifting to localized supply chains. This study estimated the economic contributions of Michigan's beer industry. The distinction of measuring the economic contribution from economic impacts is important, as an economic impact estimate would assert a change in economic activity in the absence of Michigan's brewing value chains. Rather than focus on such hypothetical scenarios, our measures detail the contributions of Michigan's beer production value chains to existing economic activities. Our findings suggest that Michigan-produced beers generate a sizeable contribution to the state economy, contributing just under $500 million to the annual Gross State Product (a measure of total income generated in the state), or alternatively, 8.4 percent of the value of Michigan's food and beverage and tobacco products manufacturing sector's annual Gross State Product [45].

Some limitations remain. These estimates omit some sources of contribution along Michigan's beer-production value chain. Namely, Michigan's agricultural producers of malting grains and hops have found markets outside of Michigan, making raw ingredient inputs an export industry for Michigan. Michigan entrepreneurs are also experimenting with specialty malting operations, with a specific focus on supplying in-state breweries, but with the potential for extending the export options along Michigan's beer production value chains. Miller [38] estimates that Michigan's agricultural inputs and malting operations contributed $8.9 million in annual transactions in 2016. Much of this value is captured in our estimates, though the value attributed to exports is not. Furthermore, not all states are likely to have as significant an economic impact from their beer value chain, as few states are as well prepared to foster the development of the industry along the extent of the value chain. Michigan is a state with a long history of growing beer inputs such as malting barley and hops, so the potential for economic impact can be relatively large.

The findings of this study highlight the ubiquitous nature of Michigan's brewing industry from that which is produced for consumption in the state and that which is produced for export to other markets. While largely comprised of small producers, in the aggregate, it comprises a measurably large component of the Michigan economy. Interest in consuming locally-sourced beer and craft beers is increasing and we believe that the industry will continue to grow, relative to the 2016 numbers used in this estimate. There are a surprisingly large number of jobs that can be tied to Michigan-sourced beer, but most of these jobs are tied to the food and drink service sector. While beer production has long been regulated at the federal and state levels, recent changes in federal law have relaxed federal oversight and increased the authority of state regulations over beer production. Recognizing that regulation is a collective choice for managing the affairs of society, regulation without a full understanding of the economic contributions can result in money and opportunities left on the table.

Author Contributions: Conceptualization, J.R.S., A.M. and S.R.M.; Data Curation, J.R.S., A.M. and S.R.M.; Formal Analysis, S.R.M., T.M. and P.H.H.; Funding acquisition, A.M., J.R.S., and S.R.M.; Project administration, A.M. Methodology, S.R.M., A.M. and J.R.S.; Writing original draft, S.R.M. and T.M.; Writing—review & editing, S.R.M., T.M., P.H.H., A.M. and J.R.S. Visualization, P.H.H.

Funding: This research was funded by a 2016 Value Add and Regional Food System Grant provided by the Michigan Department of Agriculture and Rural Development.

Acknowledgments: The authors wish to acknowledge the Michigan Department of Agricultura and Rural Development for funding this research and the Michigan Brewers' Guild for their assistance in collecting and locating the relevant information that was key to this study and for their assistance in securing funding for this project. Further, appreciation is extended to three anonymous referees, whose comments and insights greatly improved this manuscript.

Conflicts of Interest: The authors declare no conflict of interest.

References

1. Cooke, S.C.; Watson, P. A comparison of regional export enhancement and import substitution economic development strategies. *J. Reg. Anal. Policy* **2011**, *41*, 1–15.
2. Winfree, J.; Watson, P. The Welfare Economics of Buy Local. *Am. J. Agric. Econ.* **2017**, *99*, 971–987. [CrossRef]
3. Glaeser, E.L.; Gottlieb, J.D. Urban resurgence and the consumer city. *Urban Stud.* **2006**, *43*, 1275–1299. [CrossRef]
4. Glaeser, E.L.; Kolko, J.; Saiz, A. Consumer city. *J. Econ. Geogr.* **2001**, *1*, 27–50. [CrossRef]
5. Malone, T.; Lusk, J.L. Mitigating Choice Overload: An Experiment in the U.S. Beer Market. *J. Wine Econ.* **2019**, 1–23. [CrossRef]
6. Malone, T.; Lusk, J.L. Brewing up entrepreneurship: Government intervention in beer. *J. Entrep. Public Policy* **2016**, *5*, 325–342. [CrossRef]
7. Malone, T.; Chambers, D. Quantifying Federal Regulatory Burdens in the Beer Value Chain. *Agribus. Int. J.* **2017**, *33*, 466–471. [CrossRef]
8. Gohmann, S.F. Why Are There so Few Breweries in the South? *Entrep. Theory Pract.* **2016**, *40*, 1071–1092. [CrossRef]
9. Malone, T.; Hall, J.C. Can Liberalization of Local Food Marketing Channels Influence Local Economies? A Case Study of West Virginia's Craft Beer Distribution Laws. *Econ. Bus. Lett.* **2017**, *6*, 1–5. [CrossRef]
10. Malone, T.; Stack, M. What Do Beer Laws Mean for Economic Growth? *Choices* **2017**, *32*, 1–7.
11. Brewers Association. *Michigan Craft Beer Sales Statistics*; Brewers Association: Boulder, CO, USA, 2017; Available online: https://www.brewersassociation.org/statistics/by-state/ (accessed on 18 February 2019).
12. Miller, S.R. *Michigan Craft Beer: An Economic Impact Assessment of the Locally Sourced Supply Chain*; Michigan State University, Center for Economic Analysis: East Lansing, MI, USA, 2017.
13. Elzinga, K.G.; Tremblay, C.H.; Tremblay, V.J. Craft beer in the United States: History, numbers, and geography. *J. Wine Econ.* **2015**, *10*, 242–274. [CrossRef]
14. Choi, D.Y.; Stack, M.H. The all-American beer: A case of inferior standard (taste) prevailing? *Bus. Horiz.* **2005**, *48*, 79–86. [CrossRef]
15. Almenberg, J.; Dreber, A.; Goldstein, R. *Hide the Label, Hide the Difference? Working Paper No. 165*; American Association of Wine Economists: New York, NY, USA, 2014.

16. McCullough, M.; Berning, J.; Hanson, J.L. Learning by Brewing: Homebrewing Legalization and The Brewing Industry. *Contemp. Econ. Policy* **2019**, *37*, 25–39. [CrossRef]

17. Brewers Association. *Craft Brewer Defined*; Brewers Association: Boulder, CO, USA, 2017; Available online: https://www.brewersassociation.org/statistics/craft-brewer-defined/ (accessed on 29 August 2017).

18. Mintel. *Beer: US, January 2016*; Mintel Group Ltd.: London, UK, 2016.

19. Malone, T.; Lusk, J.L. If you brew it, who will come? Market segments in the US beer market. *Agribus. Int. J.* **2018**, *34*, 204–221. [CrossRef]

20. Hart, J. Drink Beer for Science: An Experiment on Consumer Preferences for Local Craft Beer. *J. Wine Econ.* **2018**, *13*, 1–13. [CrossRef]

21. Howard, P. Craftwashing in the U.S. Beer Industry. *Beverages* **2018**, *4*, 1–13. [CrossRef]

22. Vandermeulen, V.; Verspecht, A.; Vermeire, B.; Van Huylenbroeck, G.; Gellynck, X. The use of economic valuation to create public support for green infrastructure investments in urban areas. *Landsc. Urban Plan.* **2011**, *103*, 198–206. [CrossRef]

23. Bess, R.; Ambargis, Z.O. Input-output models for impact analysis: Suggestions for practitioners using RIMS II multipliers. In Proceedings of the 50th Southern Regional Science Association Conference, New Orleans, LA, USA, 24–26 March 2011.

24. Caskie, P.; Davis, J.; Moss, J.E. The economic impact of BSE: A regional perspective. *Appl. Econ.* **1999**, *31*, 1623–1630. [CrossRef]

25. Henneberry, S.R.; Whitacre, B.E.; Agustini, H.N. An evaluation of the economic impacts of Oklahoma farmers markets. *J. Food Distrib. Res.* **2009**, *40*, 64–78.

26. Jablonski, B.B.; Schmit, T.M.; Kay, D. Assessing the economic impacts of food hubs on regional economies: A framework that includes opportunity cost. *Agric. Resour. Econ. Rev.* **2016**, *45*, 143–172. [CrossRef]

27. Lillywhite, J.M.; Crawford, T.L.; Libbin, J.; Peach, J. *New Mexico's Pecan Industry: Estimated Impacts on the State's Economy*; New Mexico State University, Agricultural Experiment Station: Las Cruces, NM, USA, 2007.

28. Miller, S.R.; Mann, J.; Barry, J.; Kalchik, T.; Pirog, R.; Hamm, M.W. A replicable model for valuing local food systems. *J. Agric. Appl. Econ.* **2015**, *47*, 441–461. [CrossRef]

29. Rossi, J.D.; Johnson, T.G.; Hendrickson, M. The economic impacts of local and conventional food sales. *J. Agric. Appl. Econ.* **2017**, *49*, 555–570. [CrossRef]

30. Watson, P.; Cooke, S.; Kay, D.; Alward, G.; Morales, A. A method for evaluating the economic contribution of a local food system. *J. Agric. Resour. Econ.* **2017**, *42*, 180–194.

31. Baumol, W.J. Leontief's great leap forward: Beyond Quesnay, Marx and von Bortkiewicz. *Econ. Syst. Res.* **2000**, *12*, 141–152. [CrossRef]

32. Coughlin, C.C.; Mandelbaum, T.B. A consumer's guide to regional economic multipliers. *Fed. Reserve Bank St. Louis Rev.* **1991**, *73*. [CrossRef]

33. Watson, P.; Wilson, J.; Thilmany, D.D.; Winter, S. Determining economic contributions and impacts: What is the difference and why do we care. *J. Reg. Anal. Policy* **2007**, *37*, 140–146.

34. IMPLAN Group Llc. *Implan Version 3.0 Trade Flow Estimation*; IMPLAN Group Llc: Huntersville, NC, USA, 2012.

35. Ramey, V.A. Can government purchases stimulate the economy? *J. Econ. Lit.* **2011**, *49*, 673–685. [CrossRef]

36. Sorini, M. Understanding the Three-Tier System: Its Impacts on U.S. Craft Beer and You. CraftBeer.com. 2017. Available online: https://www.craftbeer.com/craft-beer-muses/three-tier-system-impacts-craft-beer (accessed on 30 January 2019).

37. Turnwall, A. A Year of Data Reveals Trends in the Orders of American Barroom Patrons .... Bevspot. 2017. Available online: https://www.bevspot.com/2017/04/26/draft-vs-bottle-a-breakdown-of-beer-ordering-habits/ (accessed on 1 September 2017).

38. Miller, S.R. *Michigan Beer: An Economic Contribution Assessment of Michigan's Value Chain*; Michigan State University, Center for Economic Analysis: East Lansing, MI, USA, 2017.

39. Sirrine, R.; Suttons Bay, MI, USA. Personal Communications, 2016.

40. Malone, T.; Lusk, J.L. An Instrumental Variable Approach to Distinguishing Perceptions from Preferences for Beer Brands. *Manag. Decis. Econ.* **2018**, *39*, 403–417. [CrossRef]

41. Jackson, R.W. Regionalizing National Commodity-by-Industry Accounts. *Econ. Syst. Res.* **1998**, *10*, 223–238. [CrossRef]

42. Statistica. The Countires Spending the Most on Beer. 2018. Available online: https://www.statista.com/chart/12508/the-countries-spending-the-most-on-beer/ (accessed on 5 February 2019).

43. U.S. Census Bureau. Annual Estimates of the Resident Population for the United States, Regions, States, and Puerto Rico: April 1, 2010 to July 1, 2018. Available online: https://www.census.gov/data/tables/time-series/demo/popest/2010s-state-total.html#par_textimage_1574439295 (accessed on 5 February 2019).
44. Mintel. *Craft Beer: US, October 2016*; Mintel Group Ltd.: London, UK, 2016.
45. Bureau of Economic Analysis. Table SAGDP2N. 2019. Available online: https://www.bea.gov/data/economic-accounts/regional (accessed on 11 March 2019).

Article

Sorghum Coffee–Lactose Stout Production and Its Physico-Chemical Characterization

Fali Mbeh. Harry [1], Desobgo Zangué Steve Carly [2],* and Nso Emmanuel Jong [1]

[1] Department of Process Engineering, National School of Agro-Industrial Sciences (ENSAI) of the University of Ngaoundere, P.O. Box 455 ENSAI, Ngaoundere, Cameroon; falimbeh@yahoo.com (F.M.H.); nso_emmanuel@yahoo.fr (N.E.J.)

[2] Department of Food Processing and Quality Control, University Institute of Technology (UIT), University of Ngaoundere, P.O. Box 455 UIT, Ngaoundere, Cameroon

* Correspondence: desobgo.zangue@gmail.com; Tel.: +237-697-16-00-04

Received: 23 June 2018; Accepted: 2 February 2019; Published: 1 March 2019

Abstract: Sorghum (*Safrari*) was valorized into sorghum coffee-lactose stouts using *Vernonia amygdalina* as a bittering ingredient. These sorghum grains and subsequent sorghum pale malt were tested for their acceptability in the brewing field. Results obtained were the germinative capacity of $99.29 \pm 0.58\%$, a germinative energy of $98.56 \pm 1.79\%$, a thousand corns weight 48.08 ± 0.02 g for the grains, and a diastatic power of 187.44 ± 7.89 WK for sorghum malt. The worts and beers produced were characterized and were found suitable. Moreover, alcohol content in stout beers obtained was between 8.8% and 9.4% ABV. Sensory evaluation was implemented on beers using 30 panellists and the best combination was the one using 50% lactose (250 g) and 50% coffee (250 g) in 5 L of wort during wort boiling.

Keywords: *Safrari*; coffee; lactose; stout beer; sensory evaluation

---

1. Introduction

Sorghum is Africa's fourth most important crop in terms of tonnage after maize, rice, and wheat [1]. In certain parts of Africa and India, sorghum grain has traditionally been used in the production of porridge, alcoholic beverages, and for bread making [2]. More than 300 million people in developing countries rely on sorghum as an energy source [3–5]. This is the case in Cameroon where sorghum is the largest energy source in the northern part [6–8] with a significant annual production of 1,102,000 tons [1]. Given the competition of multinational enterprises, sorghum appears to be the best alternative to lager beer production [8]. Barley has become the basic raw material for brewing for both barley and non-barley-producing countries like Cameroon. The process of brewing lager beer with cereals other than barley is growing to be a common practice, especially in non-barley-producing countries due to various drawbacks related to barley. Barley is a temperate cereal and the African climate is unsuitable for its cultivation. This presents a major problem; barley grain or barley malt must be imported, hence there are skyrocketing prices due to strong global demand and high shipping costs, meaning some valuable foreign exchange and increasing the price of the beer beyond the reach of most Africans. Importation also disadvantages local farmers as it denies them potential markets [9]. This occurs in Cameroon and these force a re-think by industries to invest in other cereals as a malting substitute. Sorghum can be well and efficiently utilized industrially in Cameroon for producing beer. The chemical composition of sorghum reassures it as an alternative cereal to barley in lager style beer production. It was previously believed that beer could not be produced without barley. However, it has been well-documented that cereals like sorghum have the potential to be an alternate substrate for conventional beer brewing. Research studies into sorghum are progressing rapidly and making a significant impact in brewing despite the earlier misunderstanding that malted sorghum developed insufficient hydrolytic enzymes [6–10]. Differences in malting and mashing temperatures employed in studies of sorghum in the past were an important contributory factor and complicated our understanding of the physiological behavior of sorghum during processing [6–10]. In recent times, the large body of work carried out on sorghum to understand the physiological behavior of sorghum has led to improved malting and mashing processes amongst other findings, such as improved varieties of sorghum suitable for beer brewing, which contributes in producing commercially acceptable sorghum beers both lager and stout [6,7,11–15]. Nigeria, South Africa, Uganda, Tanzania, Zambia, and Zimbabwe successfully brew commercial sorghum lager and stout beer [16]. The utilization of a locally grown crops as a brewing material is cost effective and can potentially boost the regional economy of Cameroon. It would benefit sorghum farmers with guaranteed income and thus reduce unemployment. In addition, the use of a native crop would reduce logistical costs for manufacturers resulting in reduced beer retail prices [17]. Manufacturers would be able to negotiate lower taxes with the government on sorghum-based beer, which would contribute to another significant cost reduction of particular benefit to the consumer. It would help create affordable lager beer for consumers for whom this type of beer is unaffordable. These factors would thus lead to an increased growth in the brewing industry [17]. Sorghum beer is gluten-free and can be used by celiacs [18,19]. In order to valorize sorghum, we came up with producing a coffee–lactose sorghum stout. The choice of coffee and lactose is not trivial. Beer customers showed interest in such a beer as they showed interest to give their impression. Bitter leaf (*Vernonia amygdalina*) was used as a substitute for hop. Bitter leaf has successfully served as a substitute to hop in lager beer. Its properties render it suitable for use [20–24]. We were interested in this paper to present how the production of the coffee–lactose stout from the malting process through and fermentation was undertaken. Thereafter, the results of the analyses conducted on worts and beers are discussed, ending up with discussing the sensory evaluation of the beers.

2. Materials and Methods

2.1. Acquisition of Materials

Safrari sorghum cultivar (Figure 1) used was obtained from the Institute of Research and Agronomic Development (IRAD), Maroua, Cameroon. The *Vernonia amygdalina* leaves (Figure 2) were harvested from farms in the Bini-dang neighborhood of Ngaoundere, Cameroon. *Coffea arabica* (Figure 3) beans used during this work were bought from a local cooperative society in Bafoussam, Cameroon.

Figure 1. *Safrari* cultivar.

Figure 2. *Vernonia amygdalina* (bitter leaf).

Figure 3. Green coffee bean.

The D-Lactose monohydrate was obtained from Sigma-Aldrich, Johannesburg, South Africa. *Saccharomyces cerevisiae* (Safbrew T-58) used for fermentation was obtained from "Malterie du Chateau", Chemin du Couloury 1, 4800 Lambermont, Belgium. The characteristics of the commercial mashing enzymes used are: Hitempase 2XL®, a thermostable α-amylase from *Bacillus licheniformis*, and Bioglucanase TX, from an enzymatic composition of β-glucanase and hemicellulases from *Trichoderma reesei* and their sources, are presented in Table 1. Hitempase 2XL and Bioglucanase TX were obtained from Kerry Bioscience, Kilnagleary, Carrigaline, Co. Cork, Ireland.

Table 1. Characteristics of commercial enzyme preparations used during mashing.

	Commercial Mashing Enzyme	
	Hitempase 2XL	**Bioglucanase TX**
Organism of origin	*Bacillus licheniformis*	*Trichoderma reesei*
Activity	4416.29 ± 19.34 U/mL	750 BGU/mL
Description	α-amylase	β-glucanase
Optimum temperature (°C)	60–95	60
Optimum pH	4–8	4.5–6.5
Recommended application level in adjuncts	60 U/g	0.01 and 0.025% (v/w)
Form	Solution	Solution

2.2. General Work Overview

Safrari cultivar was sorted to obtain homogenous samples free from foreign materials, and broken and infected grains. The sorted grains were submitted to tests of acceptability and brewing potential, which included: the germinative capacity, germinative energy, moisture content, thousand corns weight, and total ash. The grains were then malted (Figure 4) and different malts were produced by varying the kilning program. Diastatic power of the pale malt was also determined. Malted grains were mashed with a supplement of exogenous enzymes (Hitempase 2XL and Bioglucanase TX). Coffee and D-Lactose monohydrate (lactose) were added during the boiling process of wort following a mixture design to produce worts of different formulations. Dry bitter leaves (*Vernonia amygdalina*) were used as a substitute to hops. Worts produced were fermented with *Saccharomyces cerevisiae* (Safbrew T-58) at ambient temperature (25 °C). Formulations of worts, as well as the corresponding beers produced, underwent physicochemical analyses. A sensory analysis of the beers was performed. It consisted of a hedonic nine-point verbal scale with 30 panellists.

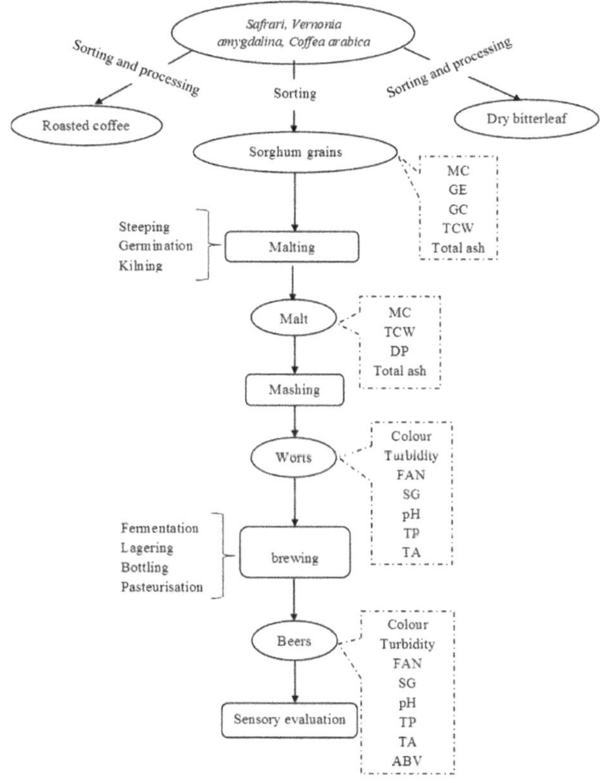

Figure 4. General work overview (MC = moisture content; GE = germinative energy; GC = germinative capacity; TCW = thousand corn weight; DP = diastatic power; FAN = free amino nitrogen: SG = specific gravity; TP = total polyphenols; TA = titratable acidity; ABV = alcohol by volume).

2.3. Processing of Raw Bitter Leaves (Vernonia Amygdalina)

As showed in Figure 5, the leaves were screened to remove foreign bodies and withered leaves. They were thoroughly washed afterwards with tap water, rinsed, and oven-dried at 30 °C for 9 days. Dried leaves were hand crushed and conditioned in glass bottles, in a refrigerator at 4 °C.

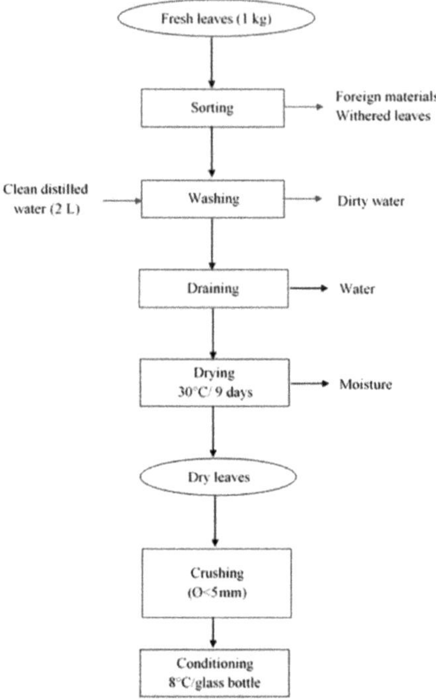

Figure 5. Processing of *Vernonia amygdalina* leaves.

2.4. Processing of Coffee Beans

The coffee beans were bought with their parchment. They were thus dehulled manually to obtain the green beans. The green beans were roasted at 200 °C for 20 min in a Memmert ventilated oven (ULM/SLM 400–800), Germany. They were subsequently cooled before being ground and conditioned in glass bottles. Figure 6 depicts the process used.

2.5. Preparation of Sorghum Grains

Sorghum grains were cleaned by first manually sorting to remove deformed, small, broken, and immature kernels, dust, sand, stones, and other foreign materials.

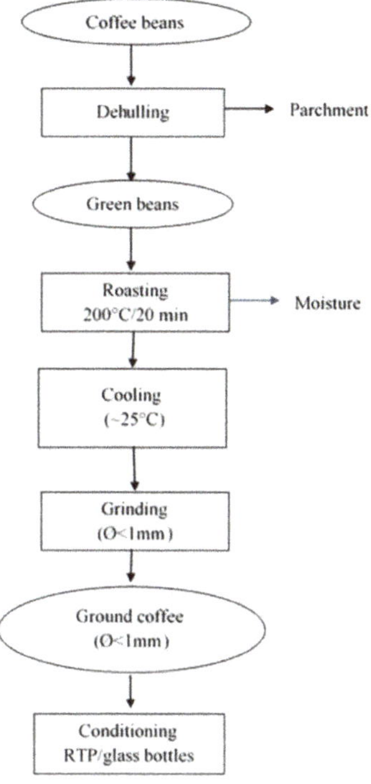

Figure 6. Processing of coffee beans.

2.6. Analysis of Sorghum Grain

2.6.1. Determination of the Germinative Energy of Sorghum

Two filter papers were placed at the bottom of two Petri dishes. A total of 4 mL and 8 mL of distilled water were accurately added [25].

Two lots of 100 corns were placed on the paper so that each made good contact without drowning the embryo by ensuring that the ventral side only touched the paper. After which, the dish was covered to prevent moisture loss, and placed in a dark cupboard. Chitted corns were removed after 24, 48, and 72 h from the beginning of steeping, thus avoiding excessive moisture uptake by those corns that germinated early. The germinative energy was calculated using the formula:

$$GE = 100 - N_{ng} \tag{1}$$

where GE is the germinative energy (%) and N_{ng} is the number of grains that had not chitted.

2.6.2. Determination of the Germinative Capacity of Sorghum

Three lots of 200 corns were obtained after removing foreign matter and half of the corns out of each of the lots of 200 corns was steeped in 200 mL of fresh H_2O_2 solution at a concentration of 7.5 g/L at room temperature (22–25 °C) for 2 days. After a day, the steep liquor was drained off and replaced with 200 mL of fresh H_2O_2 solution at room temperature. After 2 days of steeping the steep liquor was drained off and corns were separated and counted for those that had not developed both roots and acrospires [25]. The germinative capacity was calculated using the formula:

$$GC = \frac{200 - N_{ng}}{2} \tag{2}$$

where GC is the germinative capacity (%) and N_{ng} is the number of grains that had not chitted.

2.6.3. Determination of the Thousand-Corn Weight of Sorghum

Three lots of 35 g of sorghum was sampled and weighed, and half of the corns and foreign matter was removed and the weight subtracted, after which the corns were counted in each lot. This experiment was repeated thrice to obtain a more precise and accurate value. The thousand-corn weight on the dry matter was calculated using a standard formula [25]:

$$TCW = \frac{1000 \times (100 - M) \times W}{100 \times N} \tag{3}$$

where TCW is the weight of a thousand corns of dry sorghum in g, W is the weight of lot of sorghum taken in g, M is the moisture% (m/m), and N is the number of corns in each lot taken.

2.6.4. Determination of the Moisture Content

Twenty grams of sorghum was finely milled using a hand grinding machine and 5 g of the flour obtained was put in a clean dry dish and dried at of 105 °C for 24 h. The product was later removed from the oven and was immediately allowed to cool in a desiccator and weighed again. The moisture content percentage (%) of the sample was calculated using a standard formula [25]:

$$MC = \frac{100 \times (M_0 - M_1)}{M_0} \tag{4}$$

where MC is the moisture content (%), M_0 is the mass in g of the sample before drying, and M_1 is the mass in g after drying.

2.6.5. Determination of Total Ash

The sample was completely incinerated until obtaining white ash in a muffle furnace calibrated at 550 °C [26]. For that, the porcelain crucibles containing the samples resulting from drying at 105 ± 2 °C (M_2) were placed in the furnace. After incineration for 24 h, the crucibles were removed from the furnace by using grips, then cooled in the atmosphere of a desiccator and weighed (M_3). The ash content per 100 g of DM (dry matter) was calculated using the formula:

$$Total\ ash = \frac{100 \times (M_3 - M_1)}{M_2} \tag{5}$$

where M_1 is the mass of the empty crucible.

2.7. Experimental Procedure for Malting Sorghum

Seven kilograms of *Safrari* sorghum cultivar grains were washed three times using 21 L of distilled water to remove dirt and foreign bodies. Grains were steeped in 21 L of distilled water for 48 h at room temperature ($\approx$25 °C) with three changes of water at intervals of 12 h before steep out. Germination was carried out for 4 days in a Heraeus type incubator (D-63450 Hanau, Germany) at a temperature of 25 °C with water sprinkled on the grains on a daily basis. The green malt was then kilned following different kilning programs to obtain characteristic malts used in the beer recipe. The malt was rubbed-off of its rootlets and stored until further use. Different malts were produced by varying the temperature and time for kilning the green malt [27]. Four malts were used in the course of this work: base malt, caramel malt, toasted malt, and roasted malt. Figure 7 shows the malting process.

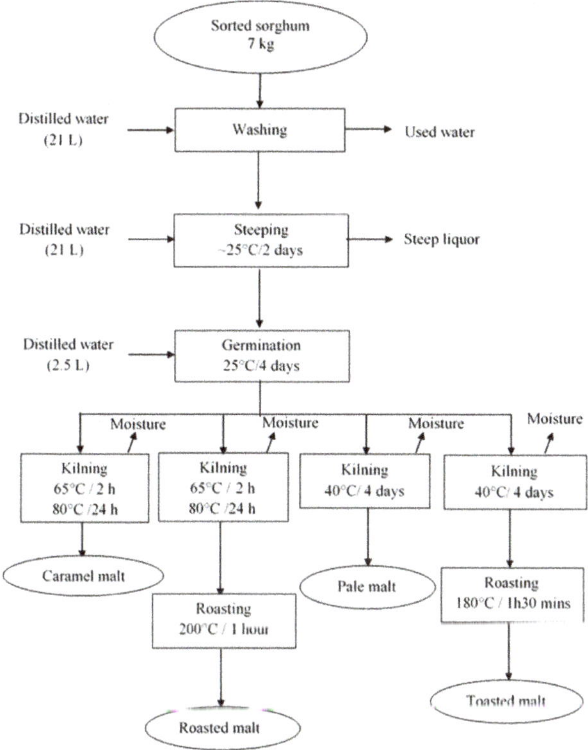

Figure 7. Sorghum malting process.

2.8. Determination of the Diastatic Power of Pale Malt

For the enzyme extraction, a water bath was first set at 40 °C. Ten grams of flour was measured into a beaker and 480 mL of water added. The mixture was mixed to avoid balling. The beaker was placed in the water bath and mashed for 1 h while stirring continuously. The extraction solution was cooled to room temperature. The beaker contents were adjusted to 510 g. The contents of the beaker were stirred and transferred onto a filter. The first 200 mL were discarded and the next 50 mL immediately used for analysis. A hundred mililiters of starch solution was pipetted into a 200 mL volumetric flask. Five mililiters of sodium acetate buffer was added and the flask placed in the water bath at 20 °C and allowed to stand for 20 min. Five mililiters of the flour extract was added using a pipette, then the contents of the flask was shaken thoroughly and replaced in the bath for 30 min. Sodium hydroxide (4 mL) was added to the mix to inactivate the enzymes. The volume was made up to 200 mL with water and mixed well. The alkalinity of the solution was verified using a thymolphthalein solution. The blank was prepared by pipetting 100 mL of starch solution into a 200 mL volumetric flask. NaOH solution (2.35 mL) was then added and mixed thoroughly. The malt extract (5 mL) was then added and the volume made up to 200 mL. The alkalinity was also checked using the thymolphthalein solution. The determination of the reducing sugars was done by transferring a 50 mL aliquot of the digest into a 150 mL Erlenmeyer flask. Into this flask was added 25 mL of iodine solution and 3 mL of sodium hydroxide, and was shaken. The flask was stoppered and allowed to stand for 15 min. Sulfuric acid (4.5 mL) was added and the unreacted iodine titrated with thiosulphate solution until the blue colour disappeared [25].

$$DP_1 = F \times (V_B - V_T)$$ (6)

$$DP_2 = \frac{100 \times DP_1}{100 - M}$$ (7)

where, DP_1 is the diastatic power on sample in Windisch–Kolbach (WK) units; DP_2 is the diastatic power on dry malt in Windisch–Kolbach units; V_B is the titration value of the unreacted iodine in the blank test, in mL; V_T is the titration value of unreacted iodine in the test portion, in mL; F is the correction factor to obtain the result per 100g of flour used for the extraction; and M is the moisture content of the flour in percentage (m/m).

2.9. Experimental Procedure for Mashing Sorghum

A decantation mashing program [7] for sorghum was adopted for the study (Figure 8). The quantities and types of malts used are given in Table 2 below. It was done according to preliminary studies.

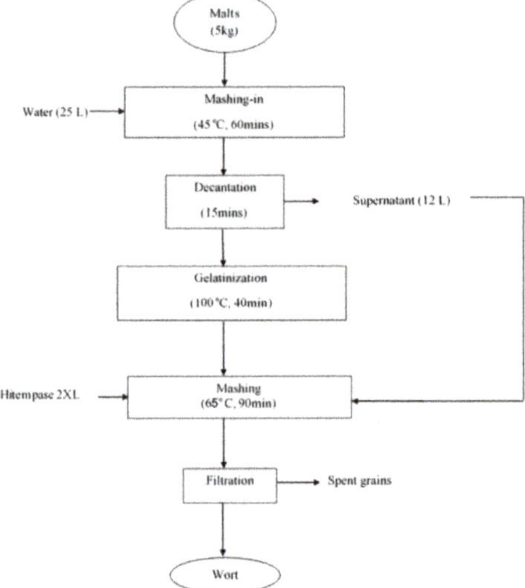

Figure 8. Decantation mashing process.

Table 2. Sorghum malt types and quantities used during mashing.

Sorghum Malt Type	Quantity/kg
Pale malt	3.1
Caramel malt	1.2
Toasted malt	0.3
Roasted malt	0.4

Twenty-five litres of distilled water was put into a stainless-steel pot and 5 kg of a mixture of sorghum malts (flour $\varnothing < 1$ mm) as given in Table 2, added with continuous stirring until a homogenous mixture was obtained. This mixture was kept at 45 °C for 1 h with intermittent stirring at intervals of 5 min. The mixture was allowed to decant and 12 L of the supernatant was withdrawn and kept aside. The temperature of the mash was then raised to boiling so as to gelatinize the sorghum starch. This was done for 40 min with intermittent stirring at intervals of 5 min before cooling to 65 °C. The supernatant withdrawn together with the commercial enzymes (Hitempase 2XL) were added to the mash and held at 65 °C for 1 h 30 min with intermittent stirring at intervals of 10 min. The mash was filtered at 25 °C using Whatmann no. 4 paper to obtain the sweet wort.

2.10. Wort Boiling

The motherwort was divided into five portions of 5 L. Each portion (5 L) of wort was boiled for an hour. During boiling, 35 g of dry bitter leaves per 5 L was added as a substitute for hops. Coffee and lactose (lactose) in a proportion of 10% (w/v) was also added 5 min before the end of wort boiling following a mixture design generated by the software Design Expert ® Version 7.0.0 (Stat-Ease, Inc. 2021 East Hennepin Ave., Suite 480 Minneapolis, MN 55413) as showed in Table 3 below.

Table 3. Experimental matrix.

Std	Run	Block	Proportion		Quantity (g)	
			x_1: Lactose	x_2: Coffee	x_1: Lactose	x_2: Coffee
4	1	Block 1	1.000	0.000	500	0
1	2	Block 1	0.750	0.250	375	125
3	3	Block 1	0.500	0.500	250	250
2	4	Block 1	0.250	0.750	125	375
5	5	Block 1	0.000	1.000	0	500

2.11. Clarification and Fermentation

The bitter worts were then cooled to room temperature, filtered to clarify them, characterized, pitched, and then fermented at room temperature for 9 days. Figure 9 illustrates the fermentation scheme.

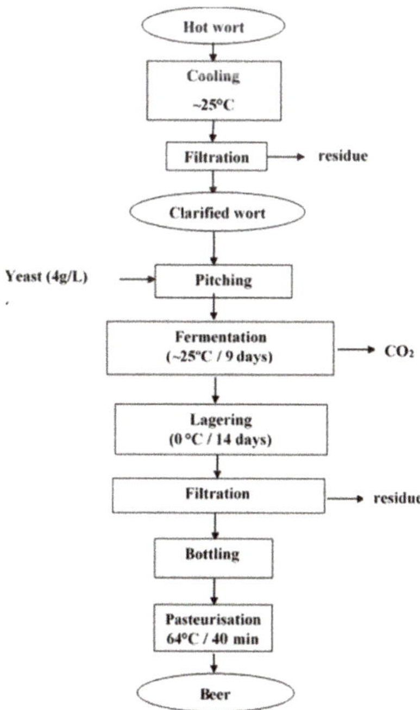

Figure 9. Brewing process.

2.12. Physicochemical Analyses of Worts and Beers

2.12.1. Determination of Specific Gravity (Analytica-EBC, 1998)

A thoroughly cleaned pycnometer was washed and rinsed with distilled water. It was then dried in an oven set at 105 °C for 5 min and then cooled in a desiccator. The initial mass of the pycnometer was weighed and then filled with the sample (wort, beer). The difference in mass between the empty pycnometer and when it was filled with the sample was evaluated. This mass was used to calculate the specific gravity by dividing the value obtained by 25 mL, which corresponds to the volume of the calibrated pycnometer. The specific gravity thus calculated was checked across an extract table for the corresponding concentration in °Brix and °Plato. Results from tables are expressed as °Plato using the Goldiner and Klemann table.

2.12.2. Determination of Color Using Spectrophotometric Method

The wavelength was set at 430 nm. The cell was filled with water and the absorbance set to read 0.00. The cell was then rinsed and filled with the sample (wort and beer) and the absorbance read [25].

$$Color\ (EBC) = A \times d \times 25 \tag{8}$$

where A is the absorbance at 430 nm in 10mm cell and d is the dilution factor

2.12.3. Determination of pH

The electrode of the pH-meter was immersed into the wort and beer samples and the pH was read. The beer was degassed before the pH was read.

2.12.4. Determination of Turbidity

The tube was filled with the sample and wiped carefully and thoroughly. It was then placed in the turbidimeter, closed, and read immediately. Results are expressed in NTU (nephelometric turbidity unit).

2.12.5. Determination of Free Amino Nitrogen (FAN) Content

The samples (wort and beer) were diluted to 1% (v/v); 2 mL of the diluted sample was placed in a test tube into which was added 1 mL of coloured reagent. Tubes were stoppered with aluminium foil, thoroughly homogenized, and placed in a water bath (95 °C) for exactly 16 min and then cooled in cold water (20 °C) for 20 min. After this time, 5 mL of dilution solution was added, mixed, and the absorbance was read at 570 nm against a reference sample prepared from the reagents plus 2 mL of distilled water in place of diluted wort and beer [25]. The free amino nitrogen (FAN) was calculated according to the relation:

$$FAN\ (mg/L) = \frac{A_1 \times 2 \times d}{A_2} \tag{9}$$

where A_1 is the absorbance of test solution at 570 nm, A_2 is the mean absorbance of standard solutions at 570 nm, and d is the dilution factor.

2.12.6. Determination of Titratable Acidity

Titratable acidity was determined according to the standardized method, i.e., AFNOR (1982), with 0.1 N sodium hydroxide (NaOH) in the presence of phenolphthalein indicator. Ten milliliters of the sample was pipetted into a conical flask and 0.1 mL of phenolphthalein (0.05%) was added. Titration was stopped when the initial color changed to pink and persisted for at least 30 s. The burette reading was noted. The titratable acidity (TA) is expressed in g/L H_2T:

$$TA = 75 \times N \times \frac{V}{T} \tag{10}$$

where V is the volume (mL) of the sodium hydroxide noted at endpoint, N is the concentration of the base, and T is the volume of titre.

2.12.7. Determination of Total Polyphenols

After haven prepared the reagents needed for this experiment, with the help of standard solutions of gallic acid, a scale of standardization was prepared and the samples are titrated.

The test tubes were again agitated and allowed to rest for 2 h at room temperature, after which their relative absorbance was read at 725 nm. The polyphenol mg/L content was obtained using the relationship given in the standard curve after plotting optical density against concentration [28].

2.12.8. Alcohol Determination Using Specific Gravity

The original gravity (OG) and final gravity (FG) were determined using a pycnometer as previously described in Section 2.12.1. The ABV was expressed as follows:

$$ABV \ (\% \ v/v) = \frac{(OG - FG)}{0.0075} \tag{11}$$

2.12.9. Sensory Evaluation of Beers

A hedonic test was conducted. Sensory evaluation was conducted by 30 ordinary consumers (20 males and 10 females) from the town of Ngaoundere. The consumers were selected from different age groups (21–35 years old). The prerequisites for participating in the study were that the individual consumed beer and showed an interest in participating in all test sessions. Evaluations were carried out at the Food Engineering and Technology Laboratory; the beer samples were served in a random way at temperatures of about 8 °C. All the beer samples were coded. Each participant received a series of five beers (30 mL of each formulation) served in opaque cups except when rating appearance, during which they were served in glasses. The degree of liking was rated using a nine-point hedonic scale for five main attributes i.e., smell, taste, bitterness, mouthfeel, and appearance; They were also asked to rate their overall liking for each beer. For each sample, participants were instructed to drink and swallow the beer when rating taste. Consumers were asked to drink mineral water to clean their mouth between tastings to avoid cross-contamination between samples. Also, they were asked not to smoke, eat, or drink anything, except water, 1 h before the tasting session. The panellists were invited to comment on the beers especially with respect to the attributes rated. After tasting, a purchase intent score sheet was filled by all the tasters.

3. Results

3.1. Physicochemical Characteristics of Unmalted and Malted Sorghum

The quality of beer was dictated by the nature of the raw materials. Sorghum (*Safrari*) utilized for this study was characterized in order to assess its brewing potentials. These tests (for acceptability in brewing) helped in determining the potentiality of the *Safrari* cultivar. Table 4 shows the results obtained for characterization of unmalted and malted sorghum.

Table 4. Physicochemical characteristics of unmalted and malted sorghum.

Characteristics	Unmalted *Safrari*	*Safrari* Malt
Water content (%)	8.50 ± 0.01	4.80 ± 0.56
Germinative capacity (%)	99.30 ± 0.58	N.D.
Germinative energy (4 mL) (%)	98.60 ± 1.79	N.D.
Germinative energy (8 mL) (%)	97.00 ± 1.21	N.D.
Thousand corn weight (g)	48.10 ± 0.02	38.00 ± 1.35
Diastatic power (WK)	N.D.	187.40 ± 7.89
Total ash (%)	1.30 ± 0.10	0.90 ± 0.22

N.D. = not determined.

3.1.1. Water Content

The values obtained for the water content of unmalted and malted sorghum were 8.50 ± 0.01% and 4.86 ± 0.56%, respectively (Table 4). In both cases, these values were in the range of those reported by the literature [29,30]. The samples produced were therefore considered to be suitable for storage since it was also reported that grains at a moisture content of up to 11.7% will keep safe without deterioration during storage [31].

3.1.2. Germinative Capacity and Energy

Results obtained for germinative capacity, germinative energy (4 mL), and germinative energy (8 mL), were 99.29 ± 0.58%, 98.56 ± 1.79%, and 97.00 ± 1.21%, respectively (Table 4). They all fell within the specifications of at least 95% [25]. These two properties had a direct bearing on the suitability of cereals for malting since it was mentioned in the literature that a GE $\geq$ 90% is appropriate for malting and brewing [31,32].

3.1.3. Thousand Corn Weight

The thousand-corn weight of unmalted and malted sorghum *Safrari* cultivar were 48.08 ± 0.02 g and 38.05 ± 1.35 g, respectively (Table 4). The literature stated a range of 7–61 g [7,15,33]. The samples, therefore, were one more time suitable for brewing.

3.1.4. Diastatic Power

The diastatic power (DP) of *Safrari* malt was 187.44 ± 7.89 WK (Table 4), less than the average value of 250 WK recorded in barley. This indicated a relatively low enzymatic activity, and thereafter, showed insufficient production of enzymes during malting when compared to barley. These relatively low levels of DP in sorghum may indicate the necessity of the addition of exogenous starch hydrolyzing enzymes since it was stated that the most important characteristics of good malt are high enzyme levels to degrade starch and obtain high extract yield [34].

3.1.5. Ash Content

The ash content of unmalted grains was 1.35 ± 0.01% (Table 4). It fell within the range indicated in the literature, which was between 0.3 and 1.7% [35,36]. A significant decrease in ash content was observed after malting. This was due to the removal of roots and shoots after kilning. *Safrari* malt had an ash content of 0.87 ± 0.22%. Despite the loss in ash content, *Safrari* malt was still within the 0.3 to 1.7% range reported in the literature [35,36]. This is beneficial for lager brewing as the yeast needs minerals for optimum function during wort fermentation [37].

3.2. Analysis of Worts and Beers

3.2.1. Specific Gravities

The specific gravity of the worts produced a range between 1.0508 and 1.0748 (Table 5). These values fall within the range as recorded in the literature [9,10,24]. The lowest value recorded was that of the motherwort. This increase could be explained, amongst other reasons, by the evaporation of water taking place during the boiling process. This evaporation should have contributed to wort concentration; thus, specific gravity increased in the boiled wort [18]. In the beer, specific gravity between 0.9923 and 1.0040 was obtained (Table 6). The lower final gravity could be explained by the fact that sugars were used up by yeast to convert the wort into beer; hence, the final gravity will be much lower as yeasts have consumed much of the sugar, which is denser than water, and have left alcohol in its place, which is less dense than water [18].

3.2.2. pH

The pH of all the worts studied was within the range 5.41 ± 0.00 to 5.79 ± 0.01 (Table 5). All the boiled wort had pH values higher than 5.41 ± 0.00, which was that of the motherwort. This increase could be explained by the fact that the pH of wort was dependent on the residual alkalinity of the water [37]. According to the literature, sorghum wort pH should range between 5.3–6.0 for optimum brewing [3,4,38]. All the worts produced were within this range. Equally, in the beer, pH ranged between 4.47 ± 0.02 and 4.68 ± 0.03 (Table 6). The lower pH values obtained in the beers could be explained by the fact that during fermentation, the pH continued to drop as a result of the yeast cells taking in ammonium ions (which are strongly basic) and excreting organic acids including lactic acid [37].

3.2.3. Turbidity

The proportion of lactose and coffee impacted the turbidity of wort by increasing it (Table 5). This could be due to the fact that coffee contains proteins, polyphenols, and carbohydrates [39], amongst other components that induce the formation of haze [7,8]. High turbidity values were recorded for all wort samples. The wort containing 100% coffee recorded a turbidity value of 91.00 ± 0.00 NTU (Table 5). A turbidity value of 606.00 ± 0.82 NTU was obtained in the sample containing 100% lactose (Table 5). Together with polyphenols, proteins are thought to cause haze formation during wort boiling (hot break) and wort cooling (cold break). The trub in both cases was as a result of the interaction between proteins and polyphenols which forms some complex that clumps. At the end of the boil, as the wort cooled it got more and more cloudy. The clumps of polyphenol and protein in the cold break were much smaller than the hot break. Therefore, they tended to stay in suspension longer, causing the wort to be turbid [7,8]. The lower turbidity values obtained in the beer (Table 6) could be explained by the fact that during maturation, clarification of the beer took place. This was due to natural sedimentation in the cold (lagering at 0 °C) of protein and polyphenol complexes. This ensured that turbidity owing to chemical precipitation or growth of microorganisms did not occur or, in the case of chemical precipitation, did not recur when the beer was clear and stable [7,8].

Table 5. Physicochemical characteristics of worts.

Factors				Responses				
X_1	X_2	SG	pH	Turbidity (NTU)	TP (mgGAE/L)	FAN (mg/L)	Colour (EBC)	TA (g/L H_2T)
1.000	0.000	1.0748	5.79 ± 0.01	606.0 ± 0.8	416.0 ± 0.1	308.0 ± 0.2	186.40 ± 0.12	2.58 ± 0.04
0.750	0.250	1.0728	5.73 ± 0.00	375.0 ± 1.2	487.0 ± 0.1	313.0 ± 0.2	242.50 ± 2.04	2.78 ± 0.10
0.500	0.500	1.0704	5.68 ± 0.01	281.0 ± 0.5	615.0 ± 0.1	317.0 ± 0.5	192.50 ± 0.20	2.84 ± 0.01
0.250	0.750	1.0632	5.63 ± 0.00	149.0 ± 0.5	441.0 ± 0.1	339.0 ± 0.9	146.70 ± 3.00	2.83 ± 0.00
0.000	1.000	1.0592	5.41 ± 0.01	91.0 ± 0.0	442.0 ± 0.0	286.0 ± 0.2	151.30 ± 0.12	2.84 ± 0.00

Table 6. Physicochemical characteristics of beers.

Factors				Responses					
X_1	X_2	SG	pH	Turbidity (NTU)	TP (mg GAE/L)	FAN (mg/L)	Colour (EBC)	TA (g/L H_2T)	ABV (%)
1.000	0.000	1.0040	4.64 ± 0.00	120.0 ± 0.5	108.0 ± 0.6	140.0 ± 0.1	155.00 ± 0.00	4.38 ± 0.02	9.30 ± 0.00
0.750	0.250	1.0027	4.62 ± 0.01	299.0 ± 0.5	127.0 ± 0.7	131.0 ± 0.4	189.70 ± 0.20	4.37 ± 0.02	9.20 ± 0.09
0.500	0.500	0.9987	4.68 ± 0.03	141.0 ± 0.0	135.0 ± 0.4	120.0 ± 0.5	183.90 ± 0.24	4.18 ± 0.04	9.40 ± 0.01
0.250	0.750	0.9960	4.57 ± 0.00	11.0 ± 0.1	139.0 ± 0.7	112.0 ± 0.1	109.20 ± 1.18	3.97 ± 0.05	8.80 ± 0.02
0.000	1.000	0.9923	4.47 ± 0.02	6.0 ± 0.0	146.0 ± 0.6	110.0 ± 0.2	112.50 ± 0.00	3.92 ± 0.08	8.80 ± 0.02

3.2.4. Total Polyphenols

Total polyphenols ranged between 416.00 ± 0.13 and 615.00 ± 0.06 mg GAE/L (Table 5). These values were higher than the 150 mg/L to 300 mg/L range from the literature [27,33,40]. The proportion of lactose and coffee impacted the total polyphenols of beer. This could be accounted for by the fact that coffee contains polyphenols in their composition [41,42]. Total polyphenols between 108 to 146 mg GAE/L were obtained in the beer (Table 6). Before the end of fermentation and maturation, a large amount of the polyphenolic material and protein had been removed by adhesion and coagulation, and it had then sunk to the bottom of the fermenter, hence the drop in total polyphenols of beer. However, they were still present in the final beer where they contributed in determining its quality since polyphenols contribute to flavor, astringency, the perception of bitterness, haze, oxidative effects, and antioxidative effects [43–45].

3.2.5. Free Amino Nitrogen (FAN)

The FAN of wort varied from 308.00 ± 0.23 mg/L to 339.00 ± 0.93 mg/L (Table 5). This could be explained by coffee, which contains proteins [41,42]. These proteins were broken down by proteases to FAN during mashing. It has been generally agreed that at least 120 mg/L of FAN is required to support proper yeast growth during brewing, though with the high gravity brewing processes employed in most modern breweries, recommended levels are at about 150 mg/L [46]. In beers, the FAN varied from 110.00 ± 0.23 mg/L to 140.00 ± 0.14 mg/L (Table 6). During fermentation, FAN provided nutritional support to the yeast, enabling the optimal yeast growth and efficient fermentation necessary for good head retention and foam quality. This further explained the decrease in FAN recorded compared to wort [47,48].

3.2.6. Color

The color values of the worts produced were within the range 146.75 ± 3.00 EBC to 242.50 ± 2.04 EBC (Table 5). This could be due to the fact that color depends on the grain used as raw materials and bitter leaf, as well as on the processes during the brewing. Color components were produced partly in the Maillard and caramelization reactions, and partly by the oxidation of polyphenols [7,49]. The color of beer ranged between 112.50 to 242.50 EBC (Table 6). The values obtained indicated that the beer produced was in the range of stout given that the appellation of stout with reference to colour starts as from 69 EBC as indicated on the beer colour chart based on the standard reference method.

3.2.7. Titratable Acidity (TA)

The proportion of lactose and that of coffee impact on the TA of beer. This could be accounted for by the fact that both lactose and coffee contained titratable acids in their composition as reported by Fox et al. (2015) and Wang and Ho (2009), respectively. Titratable acidity varied slightly comparatively in the formulation: between 2.58 ± 0.04 and 2.84 ± 0.01 g/L H2T (Table 5). Values of the TA of beer ranged between 3.92 to 4.38 g/L tartaric acid (Table 6).

3.2.8. Alcohol by Volume (ABV) Content of Beer

The ABV of the five beers were very high (Table 6). Three of beer formulations had ABV above 9%. They include formulations 1 to 3. A formulation with higher ABV means more fermentable sugars have been converted into ethanol during fermentation. Beers can contain up to about 12.5% ABV [10,24,37,50,51].

3.3. Sensory Evaluation Results

For beer 1 containing 100% lactose, all attributes had the majority of votes located in "the liking part" of the hedonic scale (Figure 10). Mouthfeel had the highest number of votes. Smell, taste, and bitterness scored high too. This could be the reason for the high score recorded for the overall liking. However, for this beer, some panellists commented on the appearance, saying it was hazy, making it less attractive. Some also mentioned that the colour was not bad but the haziness gave a poor appearance. They suggested that the beer be more brilliant.

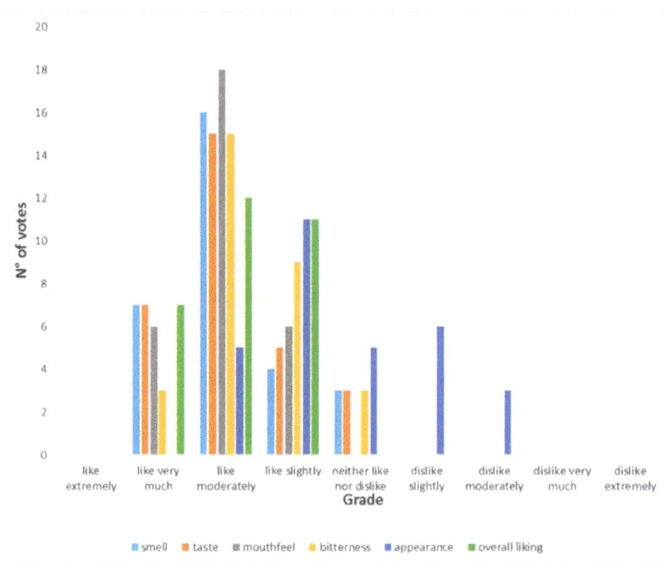

Figure 10. Hedonic nine-point scale grading for beer 1 (100% lactose and 0% coffee).

The bitterness and overall liking scored the highest in beer 2. Likewise for beer 1, for beer 2 containing 75% lactose and 25% coffee, all attributes had the majority of votes located in "the liking part" of the hedonic diagram (Figure 11). Similar comments were made by the panellists with respect to appearance being hazy. According to them, the beer lacked brilliance. They also mentioned that the colour was not bad but the haziness gave a poor appearance. They suggested that the haziness should be removed to make the beer more attractive.

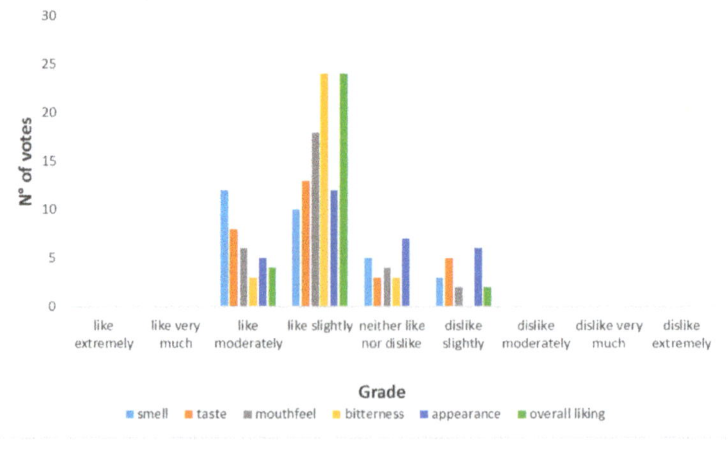

Figure 11. Hedonic nine-point scale grading for beer 2 (75% lactose and 25% coffee).

Concerning beer 3 containing equal proportions of lactose and coffee, all attributes had the majority of votes located in "the liking part" of the hedonic diagram (Figure 12). Though overall liking had the highest number of votes, a panellist said the beer could even taste better if the bitterness was reduced a little and carbonation increased. This explained the votes seen in the dislike part of the diagram. Smell, taste, and mouthfeel were, however, much appreciated in the comments made by panellists. Appearance scored higher than in the previous beer as haziness dropped.

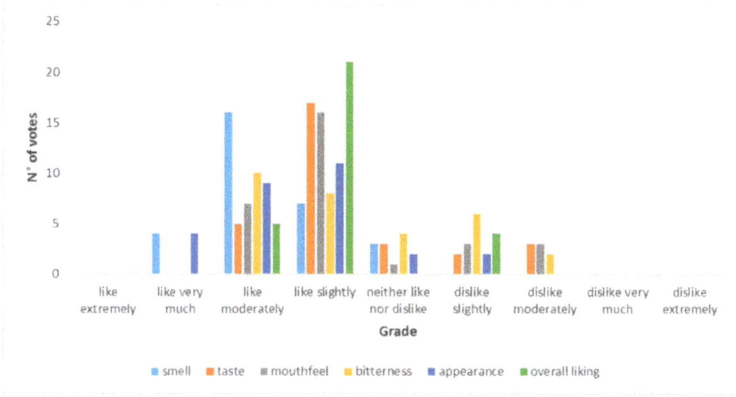

Figure 12. Hedonic nine-point scale grading for beer 3 (50% lactose and 50% coffee).

Except for smell and appearance, beer 4 containing 25% lactose and 75% coffee, had the majority of votes for all other attributes located in "the disliking part" of the diagram (Figure 13). The comments gave a clue on consumers' grading. They commented that the beer was too bitter such that they could not appreciate the flavour of the malts. They suggested much of the bitterness should be reduced. The dark colour with less haze of the beer was much appreciated and gave the appearance its good score.

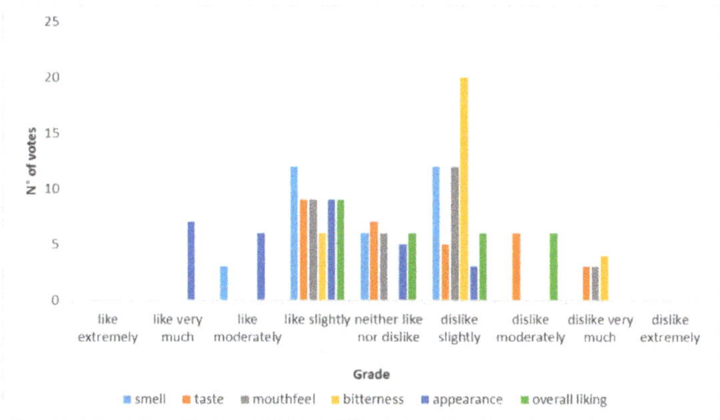

Figure 13. Hedonic nine-point scale grading for beer 4 (25% lactose and 75% coffee).

Beer 5 containing 100% coffee had the majority of votes in the "liking part" of the hedonic diagram, except for bitterness with the majority of votes located in "the disliking part" of the diagram (Figure 14). The appearance was very much liked. The dark colour with the least haze of the beer was highly appreciated as depicted by the votes. Though the bitterness was decried and suggested being reduced in the comments made by consumers, the taste scored well.

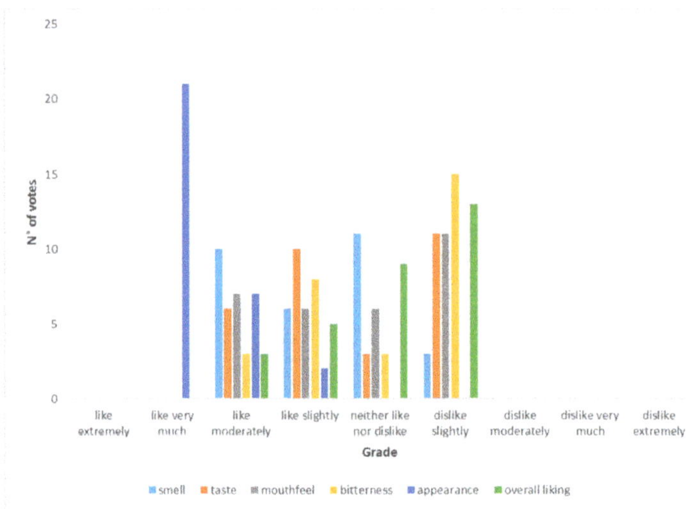

Figure 14. Hedonic nine-point scale grading for beer 5 (0% lactose and 100% coffee).

4. Conclusions

Studies on the physicochemical characterization of a beer stout using coffee and lactose as ingredients and *Vernonia amygdalina* as bittering were made. It emerged from this study after having carried out a mixing plan having given five different beers that the physicochemical characteristics of musts and beers were comparable to the values obtained in the literature. The sensory analysis of these beers revealed those which are appreciated by the panellists during a hedonic test. It is therefore at the end of this work, plausible to consider a feasibility study for a pilot production of this type of beer.

Author Contributions: F.M.H. Performed the experiments. D.Z.S.C. Analyzed the data; Contributed reagents/materials/analysis tools; Wrote the paper. N.E.J. Contributed reagents/materials/analysis tools; Supervised the laboratory work and the paper.

Funding: This research received no external funding.

Acknowledgments: The authors gratefully acknowledge the Departments of Process Engineering of the National School of Agro-Industrial Sciences (ENSAI), The University of Ngaoundere (Cameroon) for providing necessary facilities for the successful completion of this research work.

Conflicts of Interest: The authors declare no conflict of interest.

References

1. Food and Agriculture Organization of the United Nations (FAO). *Faostat Database Collections*; Food and Agriculture Organization of the United Nations: Rome, Italy, 2014.
2. Agu, R.C.; Palmer, G.H. Enzymic modification of endosperm of barley and sorghum of similar total nitrogen. *Brew. Dig.* **1998**, *73*, 30–35.
3. Bajomo, M.F.; Young, T.W. The properties, composition and fermentabilities of worts made from 100% raw sorghum and commercial enzymes. *J. Inst. Brew.* **1993**, *99*, 153–158. [CrossRef]
4. Bajomo, M.F.; Young, T.W. Fermentation of worts made form 100% raw sorghum and enzymes. *J. Inst. Brew.* **1994**, *100*, 79–84. [CrossRef]
5. Taylor, J.R.N.; Robbins, D. Factors affecting beta-amylase activity in sorghum malt. *J. Inst. Brew.* **1993**, *99*, 413–416. [CrossRef]
6. Nso, E.J.; Nanadoum, M.; Palmer, G.H. The effect of formaldehyde on enzyme development in sorghum malts. *Tech. Q. Master Brew. Assoc. Am.* **2006**, *43*, 177–182.

7. Desobgo, Z.S.C.; Nso, E.J.; Tenin, D.; Kayem, G.J. Modelling and optimizing of mashing enzymes-effect on yield of filtrate of unmalted sorghum by use of response surface methodology. *J. Inst. Brew.* **2010**, *116*, 62–69. [CrossRef]

8. Desobgo, Z.S.C. *Modélisation de L'action des Hydrolases sur Quelques Caractéristiques Physico-Chimiques des Moûts de Deux Cultivars de Sorgho*; Université de Ngaoundéré Ngaoundéré: Ngaoundéré, Cameroon, 2012.

9. Bubacz, M.; Philip, T.; McCreanor, P.T.; Jenkins, H.E. Engineering of beer: Hard work or too much fun? In *2013 American Society for Engineering Education Southeast Conference*; American Society for Engineering Education: Cookeville, TN, USA, 2013; pp. 1–10.

10. Owuama, C.I. Brewing beer with sorghum. *J. Inst. Brew.* **1999**, *105*, 23–34. [CrossRef]

11. Desobgo, Z.S.C.; Nso, E.J.; Tenin, D. The response surface methodology as a reliable tool for evaluating the need of commercial mashing enzymes for alleviating the levels of reducing sugars of worts of malted sorghum: Case of the *Safrari* cultivar. *J. Brew. Distill.* **2011**, *2*, 5–15.

12. Desobgo, Z.S.C.; Nso, E.J.; Tenin, D. Use of response surface methodology for optimizing the action of mashing enzymes on wort reducing sugars of the *madjeru* sorghum cultivar. *Af. J. Food Sci.* **2011**, *5*, 91–99.

13. Desobgo, Z.S.C.; Nso, E.J.; Tenin, D. Modeling the action of technical mashing enzymes on extracts and free-amino nitrogen yields of the *madjeru* sorghum cultivar. *J. Brew. Distill.* **2011**, *2*, 29–44.

14. Desobgo, Z.S.C.; Nso, E.J.; Tenin, D. Optimisation of the action of commercial mashing enzymes on wort extracts and free amino nitrogen of the *Safrari* sorghum cultivar. *Tech. Q. Master Brew. Assoc. Am.* **2011**, *48*, 77–86.

15. Nso, E.J.; Ajebesone, P.E.; Mbofung, C.M.; Palmer, G.H. Properties of three sorghum cultivars used for the production of *bili-bili* beverage in northern cameroon. *J. Inst. Brew.* **2003**, *109*, 245–250. [CrossRef]

16. Wolgang, K. *Technology Brewing and Malting*, 5th ed.; VLB: Berlin, Germany, 2014; p. 960.

17. Mackintosh, I.; Higgins, B. The development of a sorghum-based lager beer in uganda: A model of co-operating between industry and government in the development of local ingredients for the production of quality lager beer and consequential benefits for the parties involved. *Asp. Appl. Biol.* **2004**, *72*, 235–245.

18. Taylor, J.R.N.; Schober, J.T.; Bean, S.R. Novel food and non-food uses for sorghum and millets. *J. Cereal Sci.* **2006**, *44*, 252–271. [CrossRef]

19. Veith, K.N. *Evaluation of four Sorghum Hybrids through the Development of Glutenfree Beer*; Master of Science; Kansas State University: Manhattan, AR, USA, 2007.

20. Ajebesone, P.E.; Aina, J.O. Potential african substitutes for hops in tropical beer brewing. *J. Food Technol. Afr.* **2004**, *9*, 13–16.

21. Desobgo, Z.S.C.; Naponni, F.Y.; Nso, E.J. Characterization of *Safrari* sorghum worts and beers hopped with *Vernonia amygdalina* and nauclea diderrichii. *Int. J. Innov. Appl. Stud.* **2013**, *2*, 83–91.

22. Okafor, N. Prospects and challenges of the use of local cereals and starchy subtrates in brewing. In *National Symposium on the Brewing Industry, Now and the 21st Century*; Owerri, Nigeria, 1985; pp. 22–25.

23. Okafor, N.; Aniche, G.N. West african hop substitute for sorghum lager. *Brew. Distill. Int.* **1983**, *13*, 200–204.

24. Owuama, C.I.; Okafor, N. Use of unmalted sorghum as a brewing adjunct. *World J. Microbiol. Biotechnol.* **1990**, *6*, 318–322. [CrossRef] [PubMed]

25. Analytica-EBC. *European Brewery Convention*; Fachverlag Hans Carl: Nürnberg, Germany, 1998.

26. AFNOR. Association française de normalisation. In *Recueil des Normes Françaises des Produits Dérivés des Fruits et Légumes. Jus de Fruits*, 1st ed.; AFNOR: Paris, France, 1982.

27. Briggs, D.E. *Malts and Malting*, 1st ed.; Blackie Academic & Professional: London, UK, 1998.

28. Marigo, G. Méthode de fractionnement et d'estimation des composés phénoliques chez les végétaux. *Analysis* **1973**, *2*, 106–110.

29. Ogu, E.; Odibo, F.; Agu, R.C.; Palmer, G.H. Malting studies of some selected brewing sorghum varieties. *MBAA TQ* **2004**, *41*, 386–389.

30. Okon, E.U.; Uwaifo, A.O. Evaluation of malting sorghum i: The malting potentials of nigerian varieties of sorghum. *Brew. Dig.* **1985**, *60*, 24–27.

31. Agu, R.C.; Palmer, G.H. A reassessment of sorghum for lager-beer brewing. *Bioresour. Technol.* **1998**, *66*, 253–261. [CrossRef]

32. Agu, R.C.; Palmer, G.H. Effect of mashing with commercial enzymes on the properties of sorghum worts. *World J. Micro-Biol. Biotechnol.* **1998**, *14*, 43–48. [CrossRef]

33. Briggs, D.E.; Boulton, C.A.; Brookes, P.A.; Stevens, R. *Brewing Science and Practice*; Woodhead Publishing Limited: Cambridge, UK; CRC Press LLC: Boca Raton, FL, USA, 2004.

34. Gupta, M.; Abu-Ghannam, N.; Gallaghar, E. Barley for brewing: Characteristic changes during malting, brewing and applications of its by-products. *Compr. Rev. Food Sci. Food Saf.* **2010**, *9*, 318–328. [CrossRef]

35. Pontieri, P.; Di-Maro, A.; Tamburino, R.; De-Stefano, M.; Tilley, M.; Bean, S.R.; Roemer, E.; De-Vita, P.; Alifano, P.; Del-Giudice, L.; et al. Chemical composition of selected food-grade sorghum varieties grown under typical mediterranean conditions. *Maydica* **2010**, *55*, 139–143.

36. Pontieri, P.; De-Vita, P.; Boffa, A.; Tuinstra, M.R.; Bean, S.R.; Krishnamoorthy, G.; Miller, C.; Roemer, E.; Alifano, P.; Pignone, D.; et al. Yield and morpho-agronomical evaluation of food-grade white sorghum hybrids grown in southern italy. *J. Plant Interact.* **2012**, *7*, 341–347. [CrossRef]

37. Bamforth, C.W. *Brewing: New Technologies*; Woodhead Publishing: Cambridge, UK, 2006; p. 500.

38. Bajomo, M.F.; Young, T.W. Development of a mashing profile for the use of microbial enzymes in brewing with raw sorghum (80%) and malted barley or sorghum malt (20%). *J. Inst. Brew.* **1992**, *98*, 515–523. [CrossRef]

39. Walstra, P.; Wouters, J.; Geurts, T. *Dairy Science and Technology*; CRC Press: Boca Raton, FL, USA, 2006.

40. Briggs, D.E.; Hough, J.S. *Malting and Brewing Science*, 2nd ed.; Chapman and Hall: London, UK, 1981.

41. Fox, P.F.; Uniacke-Lowe, T.; McSweeney, P.L.H.; O'Mahony, J.A. *Dairy Chemistry and Biochemistry*, 2nd ed.; Springer International Publishing: Basel, Switzerland, 2015; p. 584.

42. Wang, Y.; Ho, C.-T. Polyphenolic chemistry of tea and coffee: A century of progress. *J. Agric. Food Chem.* **2009**, *57*, 8109–8114. [CrossRef] [PubMed]

43. Kondo, K.; Kurihara, M.; Miyata, N.; Suzuki, T.; Toyoda, M. Scavenging mechanisms of (-)-epigallocatechin gallate and (-)-epicatechin gallate on peroxyl radicals and formation of superoxide during the inhibitory action. *Free Radic. Biol. Med.* **1999**, *27*, 855–863. [CrossRef]

44. Liu, Z.; Ma, L.P.; Zhou, B.; Yang, L.; Liu, Z.L. Antioxidative effects of green tea polyphenols on free radical initiated and photosensitized peroxidation of human low density lipoprotein. *Chem. Phys. Lipids* **2000**, *106*, 53–63. [CrossRef]

45. Yilmaz, Y.; Toledo, R.T. Major flavonoids in grape seeds and skins: Antioxidant capacity of catechin, epicatechin, and gallic acid. *J. Agric. Food Chem.* **2004**, *52*, 255–260. [CrossRef] [PubMed]

46. Beckerich, R.P.; Denault, L.J. Enzymes in the preparation of beer and fuel alcohol. In *Enzymes and Their Role in Cereal Technology*; Kruger, J.E., Lineback, D., Stauffer, C.E., Eds.; American Association of Cereal Chemists: St. Paul, MN, USA, 1987; pp. 335–354.

47. Lekkas, C.; Hill, A.; Taidi, B.; Hodgson, J.; Stewart, G. The role of small wort peptides in brewing fermentations. *J. Inst. Brew.* **2009**, *115*, 134–139. [CrossRef]

48. Lekkas, C.; Stewart, G.; Hill, A.; Taidi, B.; Hodgson, J. The importance of free amino nitrogen in wort and beer. *Tech. Q.-Master Brew. Assoc. Am.* **2005**, *42*, 113.

49. Shellhammer, T.H. Beer color. In *Beer: A Quality Perspective*; Bamforth, C.W., Russell, I., Stewart, G.G., Eds.; Academic Press: London, UK, 2008; pp. 221–226.

50. Bamforth, C.W. Beer, carbohydrates and diet. *J. Inst. Brew.* **2005**, *111*, 259–264. [CrossRef]

51. Bamforth, C.W. *Chemistry of Brewing*; Elsevier Science Ltd.: Amsterdam, The Netherlands, 2003.

Article

Sensory Profile, Consumer Preference and Chemical Composition of Craft Beers from Brazil

Carmelita da Costa Jardim [1], **Daiana de Souza** [1], **Isabel Cristina Kasper Machado** [1,2], **Laura Massochin Nunes Pinto** [1], **Renata Cristina de Souza Ramos** [1] **and Juliano Garavaglia** [1,2,*]

[1] Institute of Technology in Food for Health, University of Vale do Rio dos Sinos, Av. Unisinos, 950, São Leopoldo 93022-000, Brazil; mitajardim@gmail.com (C.d.C.J.); daianasouz@unisinos.br (D.d.S.); ikasper@unisinos.br (I.C.K.M.); lauramnp@gmail.com (L.M.N.P.); rcramos@unisinos.br (R.C.d.S.R.)

[2] Department of Nutrition, Federal University of Health Sciences of Porto Alegre, Rua Sarmento Leite, 245, Porto Alegre 90050-170, Brazil

* Correspondence: julianogar@unisinos.br; Tel.: +55-51-3590-8842; Fax: +55-51-3590-8122

Received: 17 October 2018; Accepted: 14 December 2018; Published: 19 December 2018

Abstract: Craft beers are known for their distinct flavor, brew, and regional distribution. They are made using top-fermenting (ale) yeast, bottom-fermenting (lager) yeast, or through spontaneous fermentation. Craft beers are consumed and produced in Brazil in large quantities. However, they present a high level of polyphenols, which affects consumer preference as they may yield a taste of bitterness to beers. In this study, we analyzed the relationship between polyphenols and bitterness as well as the composition of the main styles of craft beers and consumer preference for them. Six different styles were analyzed according to their polyphenol content, bitterness, chemical composition, sensory profile, and preference. For preference, a panel of 62 untrained assessors was used. For sensory profile, quantitative descriptive analysis was performed using expert assessors ($n = 8$). The most preferred style was classic American pilsner, and the least preferred was standard American lager. The most preferred style showed less bitterness (9.52) and lower polyphenol content (0.61 mg EAG/mL), total solids (6.75 °Brix), and turbidity (7.27 NTU). This beer also exhibited reduced sensory notes of malty, fruity, smoked, hoppy, and phenolic but a higher perception of floral, sweet, and yeast notes; the bitterness attribute had a reduced perception. This study advances the understanding and complexity of the sensory profile of different styles of craft beers from Southern Brazil.

Keywords: craft beer; polyphenols; bitterness; preference; sensory attributes

---

1. Introduction

Beer can be defined as a product of cereal fermentation process, and it consists of more than 90% water in addition to carbohydrates, minerals, and alcohol (on average 3.5–10%) [1]. Last year, Brazil produced 13.9 billion liters and consumed 1.25 billion liters of beer, representing 7.0% and 6.6% of the global beer market, respectively [2]. Beers are primarily classified according to the fermentation process [3]. Lagers, the most consumed type of beer, are produced by low fermentation, which is usually carried out between 6 and 15 °C [2]. In contrast, ale type beers are produced by high fermentation, occurring between 16 and 24 °C after which yeast cells rise to the surface of the fermentation media, forming a thick film that is not generally removed completely [2]. Indeed, the transformation of wort into beer essentially represents the yeast-driven conversion of sugars into ethanol, CO_2, and many other secondary products that provide specific aromas and flavors [4].

Craft beer can be defined as a distinctively flavored and brewed variety that is distributed regionally. Their popularity has benefited from innovation, creativity, typicality, and authenticity, which typifies craft beer as an experience that offers pleasure, enjoyment, sense of identity and belonging, self-fulfillment, social recognition, and sustainability [5]. In recent years, there has been a big increase in the Brazilian market for craft beer consumption and production. In addition, craft beer is generally unfiltered, unpasteurized, and without additional nitrogen or carbon dioxide pressure [3]. Unlike commercial beers, craft beers are mainly produced in microbreweries following the basic brewing principles and using specific recipes according to the preference of consumers. At the same time, like commercial beers, they can be brewed using different adjuncts and yeast types [6].

It is well known that many compounds affect the sensory properties of craft beers, such as sugars, organic acids, hop bitter acids, polyphenols, and carbonyl compounds [6]. The fermentation processes through the inoculated yeast (i.e., first fermentation and refermentation in craft beers) are fundamental to the aromatic profile of the final beer produced [4]. Polyphenols are important compounds for beer quality as they can contribute to bitterness, color, body, and astringency and can therefore influence their acceptance [7,8]. Almost 67 different polyphenols have been detected in beers, both from barley and hop [9]. The most abundant phenolic acid is ferulic acid, which is found in different beer styles, especially in pilsner and weissbier [10]. Polyphenols have a key impact on the sensory quality of beers, with a higher number of polyphenols leading to better aroma and flavor of the final product [11].

Beer polyphenols come from barley malt [12] and hop [8], and their content depends on the type of beer and the quantity of hops added during its production. The brewing process and fermentation are also important factors as some chemical changes can occur during these processes [12]. Three polyphenol groups—flavan-3-ol, flavonols, and phenolic acids—are found in beers and contribute to their flavor, aroma, and chemical stability [9]. Some polyphenols act as antioxidants and prevent the oxidative degradation of beers. In addition, they provide potential benefits for human health as they inhibit mutagenic and carcinogenic agents [8].

Consumers choose craft beers because they have a variety of flavors, such as malted barley, chestnut, and honey, which increases the probability of perceiving craft beers to be of a higher quality [13]. Moreover, their consumption has become, in a qualitative approach, experienced-based and emerging from a desire for identity and distinction. The goal toward consumption is not functional but rather symbolic. [5]. Moreover, Brazilian consumers choose craft beers because they have an individual quality value and distinct sensory attributes [14]. Indeed, there has been a worldwide increase in the popularity of craft beers in recent times, particularly traditional ales, lagers, and even styles that do not fit in any of the two main types [3].

In the present study, the relationship between polyphenols and bitterness of the main styles of craft beers brewed in Southern Brazil was analyzed, along with the preference of consumers. In addition, each style of craft beer was characterized according to its chemical composition, polyphenol content, and sensory attributes. As far as we know, few researches have been conducted with the sensorial description and composition of Brazilian artisanal beer styles, evidencing the importance of this work.

2. Material and Methods

2.1. Craft Beers and Styles

Six different styles of beer were used: standard American lager (SAL), classic American pilsner (CAP), weissbier (WSB), American India pale ale (IPA), Irish red ale (IRA), and robust porter (RPO). Table 1 shows the characteristics and packaging specifications of the different craft beers. These styles were selected so that each specific beer showed different levels of color, bitterness, and ethanol content. All beer sample styles were defined according to sensory characteristics and brewing process as determined by the Beer Judge Certification Program (BJCP) [15]. The beer samples were purchased from the market and were brewed in different localities of Rio Grande do Sul State in Southern Brazil (Table 1).

Table 1. Characteristics of each craft beer samples regarding their production and packaging type. Classic American pilsner (CAP), standard American lager (SAL), weissbier (WSB), American India pale ale (IPA), Irish red ale (IRA), and robust porter (RPO).

Beer Samples	Type	Beer Color	Packing	Packing Volume (mL)	Production City	Purchase Place
CAP	Lager	Yellow	Bottle	1000	Porto Alegre	Specialty store
SAL	Lager	Yellow	Can	473	Caxias do Sul	Supermarket
WSB	Lager	Yellow	Bottle	1000	Porto Alegre	Specialty store
IPA	Ale	Red	Bottle	500	Campo Bom	Specialty store
IRA	Ale	Red	Bottle	600	Porto Alegre	Specialty store
RPO	Ale	Brown	Bottle	600	Gramado	Specialty store

2.2. Chemical Composition of Craft Beers

For all the beer parameters analyzed, the samples were decarbonated in an ultrasonic bath (Ultra Sonic Cleaner, Unique, São Paulo, Brazil) (30 min and at 80 kHz) until the foam disappeared, indicating that the beer did not contain CO_2 [16]. The turbidity was measured in a turbidity meter (TU-2016, Lutron Electronic, Taipei, Taiwan) and expressed in nephelometric turbidity units (NTU). The pH was directly measured using a calibrated pH meter (AZ 86505, AZ Instruments, Taichung City, Taiwan). The total solids were measured by refractometric method using a refractometer (Fisher Scientific, Waltham, MA, USA) and expressed in °Brix.

Dry extract was determined using an aliquot of 25 mL into metallic capsules (weighed before), evaporated in water bath for approximately 30 min, and expressed in g/L. The acidity was measured by titration with a 0.1 M NaOH solution in the presence of phenolphthalein as the indicator until the appearance of pale pink color that persisted for 1 min. The content of reducing sugar was measured using the 3,5-dinitrosalicylic acid method [17]. All procedures were carried out in triplicate, and samples were collected from the same production lot.

2.3. Beer Color

The color of craft beers was determined by colorimetric method [18,19]. The color of the beer samples was determined by HunterLAB software and a colorimeter (UltraScan PRO, Hunterlab, Reston, VA, USA) using D65 illuminating standard source, which was calibrated in the ultraviolet region for an accurate measurement of whitening agents. Aliquot of 2 mL of each craft beer was placed in a glass cell with a thickness of 2 mm. The parameters analyzed were luminosity (L*); a* (green to negative value and red to positive value); b* (blue to negative value and yellow to positive value); chroma (C*), which indicates the color purity; and angle measurement (h*), which shows the hue of the samples' color. The C* was calculated by the equation C* = $(a^{*2} + b^{*2})1/2$); the h* was measured by the equation h* = $tg^{-1}(b^*/a^*)$. Moreover, the absorbance of beer was measured at a wavelength of 430 nm in a 10-mm cuvette, and the color in European Brewing Convention (EBC) units was obtained by multiplying the absorbance by a given factor [15]. All determinations were carried out in triplicate.

2.4. Polyphenols and Antioxidant Analysis

The total phenolic content was determined using the Folin–Ciocalteu method [20]. Briefly, in 500 µL of beer samples or standard solutions, 2.5 mL of 0.2 M Folin–Ciocalteu reagent (Sigma-Aldrich, St Louis, MO, USA) and 2 mL of sodium carbonate (Sigma-Aldrich) solution (75 g/L) were added and mixed. After incubation (2 h), the absorbance was measured at 760 nm. The phenolic content was calculated from the calibration curve of Gallic acid (Sigma-Aldrich) standard solutions and expressed as millimoles of Gallic acid equivalent (GAE) per mL of craft beer. All determinations were carried out in triplicate.

The antioxidant activity was determined by 2,2-diphenyl-1-picrylhydrazyl (DPPH) radical-scavenging activity [21]. A 0.1 mL aliquot of methanolic extract was added to 3.9 mL of a 6×10^{-5} mol/L DPPH radical (Sigma-Aldrich), and the absorbance was measured at 515 nm. Quantification was performed using a calibration curve prepared with Trolox standard (6-hydroxy-2,5,7,8-tetramethylchroman-2-carboxylic acid) (Sigma-Aldrich). The results of DPPH radical-scavenging activity were expressed as µmol of Trolox per mL of beer, and all determinations were carried out in triplicate.

2.5. Determination of Bitterness

Craft beer samples were decarbonated, and bitter substances were extracted with iso-octane [16] using 10 mL of sample, 1 mL of hydrochloric acid, and 20 mL iso-octane. After this, the sample was agitated for 5 min at room temperature and then centrifuged for 15 min at 4000 rpm. The iso-octane phase was decanted and drained; the sample tube was covered and left to stand in the dark for at least 30 min before measuring the absorption at 275 nm. Results were expressed as International Bittering Units (IBU), and the average values of three determinations were used.

2.6. Sensory Analysis of Craft Beers

Ethical approval for the sensory tests of this investigation was obtained from the University of Vale do Rio dos Sinos Committee (number 12247636), and all participants gave written informed consent to participate in the study. Two different sensory tests—the ranking preference test and quantitative descriptive analysis (QDA)—was performed for each beer style.

For the ranking preference test, a hedonic panel test composed of 62 assessors who were not experienced and aged 20–56 years old was used. The selection criteria were availability and motivation to participate on all days of the experiments and the panelists being regular beer consumers. Initially, these participants answered questions about the habits of beer consumption, such as the frequency; the type, style, and brand consumed; factors that influence consumption (prize, packaging, place of consumption, etc.); sensory characteristics they appreciate the most in craft beers (aroma, flavor, color, taste, foam, etc.); and food pairing with beers. The preference was evaluated by the ranking preference test [22,23]. The test was carried out in individual cabins under white light. In each session, the beer samples were served at refrigeration temperature ranging from 6 °C to 8 °C. About 30 mL of each beer was served in transparent, glass cups without assessors having prior knowledge regarding the brand of the beer being evaluated. The samples were served randomly at the same time, and the assessors were requested to order the least preferred to the most preferred craft beers. The preference tests were carried out in four different sessions with intervals of at least eight hours between sessions to avoid sensory fatigue of the consumers. The results were submitted to Friedman test at a significance level of 5% after which the least significant difference value between the sums of the scores obtained with all analyses was calculated.

For the QDA, the flavor attributes of Southern Brazilian craft beers were analyzed [22,23]. The QDA was carried out by an experienced panel ($n = 8$) to outline the qualitative aspects of beers. Fifteen attributes, derived from literature, panelists perception, and from the attribute list used by the "beer taster association" [15], were included in the evaluation process. Seven of the attributes were related to flavor (malty, fruity aroma, floral notes, hoppy, phenolic aroma, smoked and yeast odor), two were visual attributes (foam persistency and color), five were gustatory traits (overall intensity, sweet, bitter, alcoholic, residual flavor), and one concerned the texture (level of carbonation). Industrial beers were used in pretesting panel-test sessions to let the assessors familiarize with the products under investigation and the related terminology. These sessions were also used to standardize the panel's attribute definitions according to literature and the panelists' perceptions.

The sensory attributes were assessed using an unstructured 9-point scale anchored at the left end with "absent" and at the right end with "high". The samples were identified with a code of three different random digits, where each panelist received 50 mL of each beer sample, monadically and randomly. In all sensory analysis sessions, the panelists received mineral water and dry unsalted breadsticks for palate cleansing between samples to avoid carry-over effects.

2.7. Statistical Analysis

One-way analysis of variance (ANOVA) was performed to detect statistically significant differences among the beers for the sensory attributes and chemical composition. A Tukey honestly significant difference (HSD) post-hoc test was used to identify samples that were significantly different from each other (95% significance). For ranking preference test, the Friedman test and table of Newell and MacFarlane were performed (95% significance). Statistical analysis was done using SPSS Statistics 21 software (SPSS Inc., Chicago, IL, USA). Differences of $p < 0.05$ were considered significant. Principal component analysis (PCA) was carried out on panel QDA data to identify the key attributes that most contributed to the variation in products within the product space. All PCA statistical analyses were performed with the XLSTAT, v2017 package (Addinsoft, New York, NY, USA).

3. Results

3.1. Craft Beer Composition and Color

All the craft beers tested showed best quality condition parameters according to international quality guidance [15]. Table 2 shows the composition of craft beers. In general, the craft beers had a similar composition in sugars, density, acidity, and pH; more differences were observed in turbidity, total solids, and dry extract.

The porter style (RPO) showed a higher turbidity (230 NTU) than the other tested samples. This beer had a high pH value (4.40), more solids (10% m/v), dry extract (7.47 g/L), acidity (2.19 g acetic acid/L), sugars (2.08% w/v), and ethanol (7.0% w/w). In addition, this characteristic was detected and pointed out by the hedonic panel, which described the beer as turbid and with a dark and intense color, as expected by the analysis of parameters. The SAL exhibited minor turbidity (1.44 NTU), dry extract (3.84 g/L), solids (5.75 °Brix), and acidity (1.49 g acetic acid/L).

Table 2. Principal quality parameters of each craft beer. Classic American pilsner (CAP), standard American lager (SAL), weissbier (WSB), American India pale ale (IPA), Irish red ale (IRA), and robust porter (RPO). Different letters in the same column indicate significant differences between groups of beers ($p < 0.05$, ANOVA followed by post-tests).

Style/Beer	Turbidity (NTU)	pH	Total Solids (°Brix)	Dry Extract (g/L)	Acidity (g Acetic Acid/L)	Density	Sugars (% *w/v*)	Ethanol (% *w/v*)
CAP	7.27 [e]	4.24 [c]	6.75 [b,c]	4.20 [d]	1.84 [c]	1.0112 [b]	0.9 [d,e]	5.1 [c]
SAL	1.44 [f]	4.12 [c]	5.75 [c]	3.84 [e]	1.49 [d]	1.0098 [b]	0.93 [c,d]	5.0 [c]
WSB	16.78 [d]	3.88 [d]	7 [b]	4.80 [c]	1.97 [b]	1.0116 [b]	0.95 [c]	5.0 [c]
IPA	37.77 [b]	4.12 [c]	7 [b]	4.21 [d]	1.97 [b]	1.0084 [b]	0.86 [e]	6.2 [b]
IRA	29.14 [c]	4.33 [a,b]	7.75 [b]	5.36 [b]	1.52 [d]	1.0139 [a,b]	1.13 [b]	6.2 [b]
RPO	230 [a]	4.40 [a]	10 [a]	7.47 [a]	2.19 [a]	1.0222 [a]	2.08 [a]	7.0 [a]

Regarding the color of beers, differences in L*, a*, and b* parameters were found. All samples showed high luminosity, but SAL had higher luminosity than other craft beers analyzed (Table 3). Minor L* value was detected with porter (RPO) style, a very turbid beer (Table 2). The L* value ranged from 14.02 (RPO beer) to 91.65 (SAL beer). The a* value, which represents the color axis green to red, ranged from −0.49 (SAL) to 33.43 (RPO). The positive values indicated a perception of red color due to the toasted barley use in craft beer production. For parameter b*, a tendency of yellow color was noticed and ranged from 24.03 (RPO) to 89.6 (IRA). The decrease in b* value of some samples of craft beers led to a reddish color and with a brown trace, as a function of a* value of color. Chroma value was positive for all craft beer samples and ranged from 32.74 (SAL beer) to 94.19 (IRA). The beer IRA showed a higher chroma when compared to other samples, representing a beer color with more quality, purity, and intensity. The *h* angle oscillated from −1.556 (SAL) to 1.532 (CAP), indicating a more yellow color of beer samples. The *h* is correlated to a* and b* value, and it is important to differentiate the color hue from different beer samples. The CAP beer had a more intense and yellow hue compared to the other samples (Table 3).

Table 3. Color parameters of craft beers. L* (luminosity), C* (chroma), h* (hue) and EBC (European Brewery Convention) units. Classic American pilsner (CAP), standard American lager (SAL), weissbier (WSB), American India pale ale (IPA), Irish red ale (IRA), and robust porter (RPO). Different letters in the same column indicate significant differences between groups of beers ($p < 0.05$, ANOVA followed by post-tests).

Style/Beer	L*	a*	b*	C*	h*	EBC Units
CAP	87.21 [c]	1.82 [d]	47.39 [c]	47.43 [c]	1.532 [a]	13.37 [d]
SAL	91.65 [a]	−0.49 [f]	32.73 [e]	32.74 [e]	−1.556 [e]	7.50 [e]
WSB	89.90 [b]	1.01 [e]	40.87 [d]	40.89 [d]	1.546[a]	9.75 [e]
IPA	77.12 [d]	12.13 [c]	71.72 [b]	72.74 [b]	1.403 [b]	16.75 [c]
IRA	62.46 [e]	29.06 [b]	89.60 [a]	94.19 [a]	1.257 [c]	44.75 [b]
RPO	14.02 [f]	33.43 [a]	24.03 [f]	41.17 [d]	0.623 [d]	157.0 [a]

The color expressed in EBC units varied from 7.50 (SAL) to 157 (RPO). The beer with higher EBC index (RPO: 157) showed lower luminosity (14.02), and the least intense EBC color had more luminosity (91.65).

3.2. Bitterness, Antioxidant Activity, and Polyphenols

The polyphenol content in beer is an important factor to analyze as it can improve the quality and acceptance of craft beers. Table 4 shows the content of polyphenols, antioxidant activity, and bitterness of each craft beer sample. The beers with higher level of polyphenols were RPO (1.62 mg EAG/mL), IRA (0.95 mg EAG/mL), and WSB (1.68 mg/EAL/mL). The SAL craft beer showed lower polyphenols content (0.35 mg EAG/L) compared with other samples. In Table 4, it can be seen that the beers that presented higher content of total polyphenols were also the ones with greater antioxidant activity.

The antioxidant activity was maximal (5.58 μmol Trolox/mL) with the weissbier beer (WSB) using the DPPH method. In general, the antioxidant activity of the tested beers varied from 1.74 μmol Trolox/mL (SAL) to 5.58 μmol Trolox/mL (WSB). The beer bitterness was maximal in IPA beer (46.15 EBU), and the lowest value of bitterness was 9.52 EBU (CAP) (Table 4). The bitterness EBC value varied depending on the content of bitter compounds and polyphenols in the beers. In this case, the craft beer with higher content of total polyphenols did not show a higher bitterness value.

Table 4. Total polyphenols content, antioxidant activity (2,2-diphenyl-1-picrylhydrazyl (DPPH) method), and bitterness value of different craft beers. Classic American pilsner (CAP), standard American lager (SAL), weissbier (WSB), American India pale ale (IPA), Irish red ale (IRA), and robust porter (RPO). Different letters in the same column indicate significant differences between groups of beers ($p < 0.05$, ANOVA followed by post-tests).

Style/Beer	Total Polyphenols (mg EAG/mL)	DPPH (μmol Trolox/mL)	Bitterness (IBU)
CAP	0.61 [d]	3.24 [b]	9.52 [f]
SAL	0.35 [e]	1.74 [e]	11.57 [e]
WSB	1.68 [a]	5.58 [a]	12.55 [d]
IPA	0.8 [c]	2.30 [c]	46.15 [a]
IRA	0.95 [b]	2.05 [d]	33.45 [b]
RPO	1.62 [a]	3.14 [b]	24.72 [c]

3.3. Sensory Analysis of Beers

For the hedonic test of beers, 62 panelists were recruited to evaluate six different styles. This panel was composed of 57.4% females and 42.6% males. The average consumer age was 32.09 ± 10.6 years old and ranged from 20 to 56 years old. Regarding the frequency of beer consumption, 81.5% of panel reported frequently drinking beer every day and only occasionally consuming it on a weekly basis. They also consumed both commercial and craft beer brands; the most consumed craft beers were local beers, followed by the international brands of craft beers available.

Concerning the factors that influence beer consumption, most assessors chose the beer differential and typical sensory characteristics, the type of serving, the beer label design, and the beer style. The second most important factor was the place of consumption. The least important factor was the type of packaging. The most important sensory characteristics appreciated by the survey participants were the flavor and then the beer fragrance notes. Regarding the preference for styles of beer, the most cited were pilsner, weissbier, and India pale ale.

The calories contained in beer had relevance for only eight participants (12.9%) in the survey, and the clear majority of participants said they usually drink with their friends. When talking about the consumption of beer combined with some type of gastronomic preparation, 24 people (38.7%) reported that they do not care about it, 14 (22.6%) said they do not usually drink with food, and 24 assessors (38.7%) said they try to harmonize the drink with food. Regarding the factors influencing beer consumption, a majority chose the beverage differential, such as how it is served, the label, and the style. The second most important factor was the place where they drink the beer. The least important factor was the packaging. According to sensory characteristics of craft beers, the most prominent was the taste, followed by aroma. For only eight participants (12.9%) in the survey, the calories contained in beer had relevance. On the other hand, the clear majority of participants usually drink with their friends. When talking about the consumption of beer harmonized with some type of gastronomic preparation, 20 people (31.25%) reported that they do not care about it and 14 (22.6%) do not usually drink with the food and only 20 people (31.25%) try to harmonize the drink with the food. Concerning the brewing schools (English, Belgian, German, and American), 63% did not know any of them. Regarding the preference for a particular style of beer, the most cited were pilsner, weissbier, and India pale ale.

Regarding the useful ranking preference test, the least preferred beer was IPA, and the most preferred style was pilsner (CAP). The ranking preference test was considered significant (95% significance) using the Friedman test, and there was a significant difference in the preference when comparing the scores between them. Pilsner craft beer (CAP) was more preferred when compared to lager beer (SAL) and other craft beers. In fact, none of the participants chose CAP beer as the least preferred of all beer samples. Pilsner (CAP), one of the beers with the lowest number of polyphenols (0.61 mg EAG/mL) and bitterness (9.52 IBU), had a higher preference compared to the others. Thus, an increase in polyphenol level and beer bitterness led to a decrease in preference by the panel test. The IPA beer also showed a more intense bitterness (46.15 IBU), which was a factor that contributed to its low preference among beer consumers.

Figure 1 shows the sensory profile of different craft beer styles by QDA. This data indicates the differences in the craft beer styles according to the abovementioned sensory attributes. Aroma attributes, carbonation, hoppy, and foam were some important characteristics evaluated by beer consumers.

The CAP beer showed a high sweet flavor score (3.64) but did not show high scores for other descriptors (Figure 1). The RPO beer exhibited good color (6.23), overall intensity (5.23), foam (5.34), malty (5.08), and smoked (3.48). The hoppiest (4.59) and fruitiest (5.24) craft beer was IPA. This craft beer had around 2.5-fold more hoppy flavor than CAP beer (1.8), which was the most preferred beer. The bitterest craft beer was IPA (7.40) and IRA (7.25). Consumers preferred beers without high polyphenols content, less bitterness (EBU units), and lower bitter and hoppy character.

Differences in the sensory profiles of craft beers were investigated by PCA, and the results are shown in Figure 2. This analysis matrix included all sensory attributes evaluated (Figure 1). Two principal components (PCs) were extracted, and one group of samples was discernible after analysis of PC1 versus PC2 in a biplot of samples and selected variables. In this PCA plot, PC1 explained 37.94% of total variance and PC2 explained another 29.7%. Based on the results of PCA and considering the studied beer samples, CAP, IRA, and WSB were grouped (Figure 2). The beers IPA, RPO, and SAL did not cluster together and remained separated in the plots. The group of beers showed yeast/fermentation and sweet flavor and had a high perception of carbonation. In the upper left quadrant, the beer IPA was mainly related to the presence of floral flavor. The IPA style showed a more intense perception of floral flavor and hoppy character. The RPO positioned in the upper right quadrant was more related to the presence of more intense color besides alcoholic, malty, fruity, and overall intensity attributes (Figure 2).

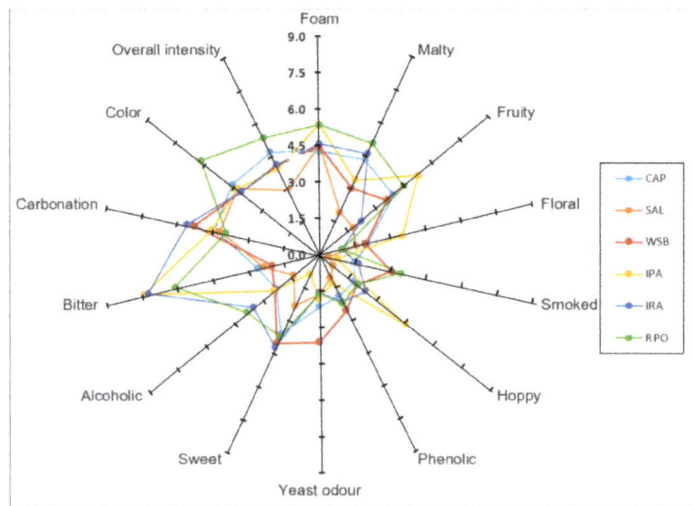

Figure 1. Plots of mean intensity scores for sensory profile of six different craft beers evaluated by quantitative descriptive analysis using a 9-point scale. Standard American lager (SAL), classic American pilsner (CAP), weissbier (WSB), American India pale ale (IPA), Irish red ale (IRA), and robust porter (RPO).

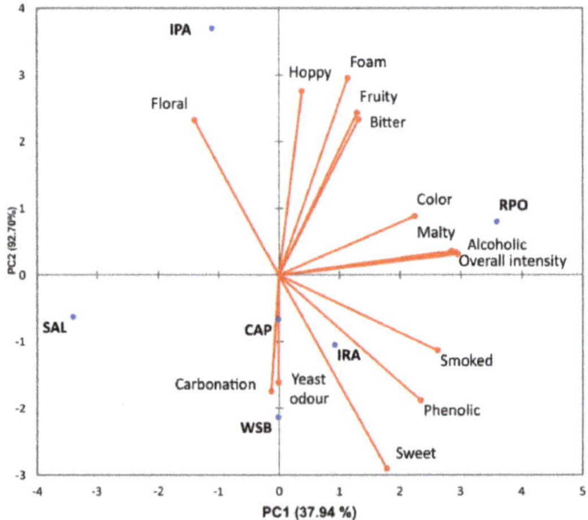

Figure 2. Scatter plots of principal component analysis (PCA) scores for specific sensory attributes of Southern Brazilian craft beers analyzed in the present study. (PC1 + PC2 explain 67.64% of total matrix variance). Standard American lager (SAL), Classic American pilsner (CAP), weissbier (WSB), American India pale ale (IPA), Irish red ale (IRA), and robust porter (RPO).

4. Discussion

Beer is a very complex mixture, and its chemical composition varies considerably [24], as shown in Table 2. To bring more light into the differences found in craft beer consumption, the objective of this work was to explore the impact of polyphenol content and bitterness of Southern Brazilian craft beers on consumer preference. As craft beers have different flavors, aromas, etc. than the usual well-known commercial brands, their preference is increasing among consumers [13].

These differences in craft beer flavors come from the ingredients used as well as the brewing process [16]. The yeast strains used have a key impact on the final craft beer quality, either in the aroma compound production or in the interaction of some beer components, such as polyphenols [4]. While several yeast strains are commercially accessible, the availability of new starter strains could be an important differentiating factor among craft beers produced in different microbreweries [25]. The main ingredients used in beer production are barley, hops, water, and yeast [26], with each ingredient playing a crucial role in the quality and composition of beers. The porter style beer, for example, is characterized as a substantial, malty dark beer with a complex and flavorful dark malt character [15]. This beer showed high scores in composition parameters compared to other beers tested in this study. Nevertheless, in general, the tested craft beers were similar in analytical factors to the styles described in the BJCP guide [15].

In addition, craft beers have distinctive and pleasant flavor characteristics, and these attributes are easily perceived by consumers [27]. Today, consumer preferences appear to be connected to the discovery of new beer flavors [13], which can increase the consumption of craft beers. Brazilian consumers have followed the same trend as they search for beers with high sensorial quality and differentiated and characteristic flavor and aroma, as verified in this study. The main factor that affects Brazilian beer consumers is the sensory attributes, as pointed out by other studies [13,27,28]. The most preferred craft beer in our study was the CAP style, which mainly shows a fruity and sweet note. Additionally, consumers have a predilection to drink with friends and consider the flavor and fragrances of beer.

It must be noted, however, that this study had some limitations, in particular the few number of craft beer samples that were evaluated for each style of beer. Even so, the sensory attributes and craft beer styles selected in this study for their consumer relevance spanned a wide range of beer characteristics.

Moreover, studying consumer behavior can have great value for the beer industry as it can show how consumers represent the beer category, the associations linked to them, and the proximity across different types of beer [27]. In addition, studies about consumer preferences can assist brewers to understand consumer attitude and translate consumer needs, wants, and expectations into manufacturing design to produce the best, most cost-competitive, and widely accepted product possible in a relatively short period [29].

Beers are rich in polyphenols, which are mostly acquired from barley and hop [8], and they were found in the six styles of beers evaluated in this study. For example, xanthohumol is the most common phenol in hop [30]. Investigating the Brazilian beers, we found that the contents of phenolic compounds, as well as the antioxidant capacity, were like those of beers produced elsewhere in the world [2]. Polyphenols are already extracted in the initial phase of the fermentation process, i.e., during the wort production [9]. This study showed that polyphenols, especially the bitterness associated with it, have an important relationship with the preference of different beers among Brazilian consumers. The most preferred beer showed the lowest bitterness (Table 4, Figure 1) of all styles tested and the second lowest level of polyphenols (0.61 mg EAG/mL). This same relationship with bitterness was verified when analyzing consumer acceptance of craft beers and commercial brands in the Brazilian market [31]. Understanding the sensory characteristic of bitterness in beers and how that relates to their content of polyphenols has a significant value for understanding consumer response as well as optimizing production processes [8].

The malt kilning process determines the color parameter, and it is quite an important process as it can improve the acceptance of beers [26]. The luminosity (L* value) also has a great importance because beers with higher L* value (high luminosity) show a more vivid and intense color [19]. The lager beers show high L* values [19].The darker and more turbid beers show big scores in fruity, floral, and malty flavors, but they still have a low preference among consumers. Beer appearance provides substantial opportunities for product differentiation, and even beers of the same type have the potential to deliver on rather different usage contexts [32].

The most popular beer style in Brazil is the Germany-style pilsner, which is very light and clear [31]. This beer style is very common in the Brazilian market and has a great familiarity to consumers. Familiar beers would be more often cited as appropriate in most of usage contexts, and that familiar and novel products would be associated with different usage contexts [32]. Consumers perceive familiar beers to be more interesting and tasty [32], which can increase their preference, as verified in this study. The preference order obtained from this study came from the sensory proprieties perceived by non-trained assessors because the beer samples were analyzed at the same time and were not assigned different styles.

From the sensory characterization of Brazilian beer styles, it is possible to attest that the evaluated consumers could differentiate and prefer the most aromatic and fruity beers. In addition, this distinct character is a motivation to choose and buy craft beers instead of other beer brands [29]. In this context, the use of *S. cerevisiae* strains isolated from food matrices can represent a valid approach for the selection of starters for brewing to obtain craft beers with more complex flavors [25]. A study with Italian consumers found similar preference to beers brewed from moderately kilned/roasted malts with a milder flavor and less intense mouthfeel perceptions [28]. More complex craft beers with remarkably increased flavors would lead to an improvement of consumer preference [29].

The IPA was the lowest preferred beer and showed a higher level of bitterness attribute perception by the panel test. According to international definitions, the IPA style is a hop-forward, bitter, dryish beer with good drinkability. However, the excessive harshness and heaviness of IPA are typical faults that result from the strong flavor clashes between the hops and other specialty ingredients [15]. Furthermore, IPA beer was differentiated by PCA (Figure 2) from other styles because of its characteristic floral note.

Bitterness is a very important quality parameter in beer production [16]. Nearly four out of ten consumers report that they highly appreciate sweet and fruity samples but dislike primarily bitter, burnt, and roasted notes and hoppy resinousness of beer [28]. The bitter foods are generally disliked due to the instinctive rejection of the bitter taste [33]. Variations in liking and the willingness to consume bitter foods can be triggered by motivational states in humans [33]. In this study, the beer with the lowest bitterness had a higher preference among consumers. This shows bitterness is a key factor that influences beer preference and leads to a decline in consumer preference.

5. Conclusions

In this study, we found that the polyphenol content and bitterness determine the preference of craft beers among Southern Brazilian as consumers can perceive their complex sensory attributes. The research also showed the influence of polyphenols in terms of consumer preference as beers with less polyphenol content and bitterness (CAP beer) were more preferred than other craft beer types. Brazilian craft beers with high antioxidant activity, polyphenols, and bitterness were the porter style (RPO), red ale (IRA) and India pale ale (IPA). The craft beers showed complex aromatic notes and flavors, which were described as floral, fruity, yeast, and malty. There were, however, some limitations in this study as it was only exploratory. Therefore, additional work with a larger craft beer sample that is representative of Brazilian craft beers is needed to strengthen our conclusion.

Considering these study findings, it is possible to describe some craft beers and point to the adverse effect of polyphenols and bitterness on consumer preferences. These preliminary results will be important in stimulating the production of more appreciable craft beers for Southern Brazilian consumers who want to improve their drinking experience and hedonic aspects.

Author Contributions: Study conception and design: All; Acquisition of data: C.d.C.J., D.d.S. and L.M.N.P.; Analysis and interpretation of data: All; Drafting of manuscript: C.d.C.J., R.C.d.S.R. and J.G.

Funding: This research was funded by Financiadora de Estudos e Projetos (FINEP) of the Brazilian Government, grant number 0110051002. Additional support was provided by Institute of Technology in Food for Health, University of Vale do Rio dos Sinos.

Acknowledgments: The authors thank the Financiadora de Estudos e Projetos (FINEP) of the Brazilian Government for financial support.

Conflicts of Interest: The authors declare no conflict of interest.

References

1. Gaetano, G.; Costanzo, S.; Di Castelnuovo, A.; Badimon, L.; Bejko, D.; Alkerwi, A.; Chiva-Blanch, G.; Estruch, R.; La Vecchia, C.; Panico, S.; et al. Effect of moderate beer consumption on health and disease: A consensus document. *Nutr. Metab. Cardiovas.* **2016**, *26*, 443–467. [CrossRef] [PubMed]
2. Moura-Nunes, N.; Brito, T.C.; Fonseca, N.D.; de Aguiar, P.F.; Monteiro, M.; Perrone, D.; Torres, A.G. Phenolic compounds of Brazilian beers from different types and styles and application of chemometrics for modeling antioxidant capacity. *Food Chem.* **2016**, *199*, 105–113. [CrossRef] [PubMed]
3. Capece, A.; Romaniello, R.; Pietrafesa, A.; Siesto, G.; Pietrafesa, R.; Zambuto, M.; Romano, P. Use of *Saccharomyces cerevisiae* var. *boulardii* in co-fermentations with *S. cerevisiae* for the production of craft beers with potential healthy value-added. *Int. J. Food Microbiol.* **2018**, *284*, 22–30. [CrossRef]
4. Budroni, M.; Zara, G.; Ciani, M.; Comitini, F. *Saccharomyces* and Non-*Saccharomyces* Starter Yeasts. In *Brewing Technology*; Kanauchi, M., Ed.; Intechopen: London, UK, 2017; pp. 81–100.
5. Gómez-Corona, C.; Lelievre-Desmas, M.; Buendía, H.E.B.; Chollet, S.; Valentin, D. Craft beer representation amongst men in two different cultures. *Food Qual. Preferences* **2016**, *53*, 19–28. [CrossRef]
6. Estela-Escalante, W.D.; Rosales-Mendoza, S.; Moscosa-Santillán, M.; González-Ramírez, J.E. Evaluation of the fermentative potential of *Candida zemplinina* yeasts for craft beer fermentation. *J. Inst. Brew.* **2016**, *122*, 530–535. [CrossRef]
7. Giacalone, D.; Ribeiro, L.M.; Frost, M.B. Perception and description of premium beers by panels with different degrees of product expertise. *Beverages* **2016**, *2*, 5. [CrossRef]
8. Oladokun, O.; Tarrega, A.; James, S.; Smart, K.; Hort, J.; Cook, D. The impact of hop bitter acid and polyphenol profiles on the perceived bitterness of beer. *Food Chem.* **2016**, *205*, 212–220. [CrossRef] [PubMed]
9. Juric, A.; Coric, N.; Odak, A.; Herceg, Z.; Tisma, M. Analysis of total polyphenols, bitterness and haze in pale and dark lager beers produced under different mashing and boiling conditions. *J. Inst. Brew.* **2015**, *121*, 541–547. [CrossRef]
10. Piazzon, A.; Forte, M.; Nardini, M. Characterization of phenolics content and antioxidant activity of different beer types. *J. Agric. Food Chem.* **2010**, *58*, 10677–10683. [CrossRef]
11. Zhao, H.; Li, H.; Sun, G.; Yang, B.; Zhao, M. Assessment of endogenous antioxidative compounds and antioxidant activities of lager beers. *J. Sci. Food Agric.* **2013**, *93*, 910–917. [CrossRef]
12. Quifer-Rada, P.; Vallverdú-Queralt, A.; Martínez-Huélamo, M.; Chiva-Blanch, G.; Jáuregui, O.; Estruch, R.; Lamuela-Raventós, R. A comprehensive characterisation of beer polyphenols by high-resolution mass spectrometry (LC-ESI-LTQ-Orbitrap-MS). *Food Chem.* **2015**, *169*, 336–343. [CrossRef] [PubMed]
13. Aquilani, B.; Laureti, T.; Poponi, S.; Secondi, L. Beer choice and consumption determinants when craft beers are tasted: An exploratory study of consumer preferences. *Food Qual. Preferencs* **2015**, *41*, 214–224. [CrossRef]
14. Ferreira, R.H.; Vasconcelos, R.L.; Judice, V.M.M.; Neves, J.T.R. Inovação na fabricação de cervejas especiais na região de Belo Horizonte. *Perspect. Ciênc. Inf.* **2011**, *16*, 171–191. [CrossRef]
15. Beer Judge Certification Program (BJCP). 2015 Style Guidelines: Beer Style Guidelines. 2015. Available online: https://www.bjcp.org/docs/2015_Guidelines_Beer.pdf (accessed on 8 November 2018).
16. Popescu, V.; Soceanu, A.; Dobrinas, S.; Stanicu, G. A study of beer bitterness loss during the various stages of the Romanian beer production process. *J. Inst. Brew.* **2013**, *119*, 111–115. [CrossRef]

17. Miller, G.L. Use of dinitrosalicylic acid reagent for determination of reducing sugar. *Anal. Chem.* **1959**, *31*, 426–428. [CrossRef]

18. Mamede, M.; Monica, S.; Maria, J.; Cruz, J.; Oliveira, L. Avaliação sensorial e colorimétrica de néctar de uva. *Braz. J. Food Nutr.* **2013**, *24*, 65–72.

19. Smedley, S.M. Discrimination between beers with small colour differences using the CIELAB colour space. *J. Inst. Brew.* **1995**, *1*, 195–201. [CrossRef]

20. Meda, A.; Lamien, C.; Romito, M.; Millogo, J.; Nacoulma, O. Determination of the total phenolic, flavonoid and proline contents in Burkina Fasan honey, as well as their radical scavenging activity. *Food Chem.* **2005**, *91*, 571–577. [CrossRef]

21. Brand-Williams, W.; Cuvelier, M.E.; Berset, C. Use of a free radical method to evaluate antioxidant activity. *Food Sci.Technol.* **1995**, *28*, 25–30. [CrossRef]

22. Dutcosky, S.D. *Análise Sensorial de Alimentos*, 3rd ed.; Editora Champagnat: Curitiba, Brazil, 2013; Chapter 4; p. 356.

23. Lawless, H.T.; Heymann, H. *Sensory Evaluation of Food: Principles and Practices*, 2nd ed.; Springer: New York, NY, USA, 2010; Chapter 4, p. 596.

24. Silva, G.C.; Abner, A.S.; Silva, L.S.N.; Godoy, R.L.O.; Nogueira, L.C.; Quitério, S.L.; Raices, R.S.L. Method development by GC–ECD and HS-SPME–GC–MS for beer volatile analysis. *Food Chem.* **2015**, *167*, 71–77. [CrossRef]

25. Marongiu, A.; Zara, G.; Legras, J.; Del Caro, A.; Mascia, I.; Fadda, C.; Budroni, M. Novel starters for old processes: Use of *Saccharomyces cerevisiae* strains isolated from artisanal sourdough for craft beer production at a brewery scale. *J. Ind. Microbiol. Biotechnol.* **2015**, *42*, 85–92. [CrossRef] [PubMed]

26. Han, H.; Kim, J.; Choi, E.; Ahn, E.; Kim, W.J. Characteristics of beer produced from Korean six-row barley with the addition of adjuncts. *J. Inst. Brew.* **2016**, *122*, 500–507. [CrossRef]

27. Gómez-Corona, C.; Valentin, D.; Escalona-Buendía, H.B.; Chollet, S. The role of gender and product consumption in the mental representation of industrial and craft beers: An exploratory study with Mexican consumers. *Food Qual. Preferences* **2017**, *60*, 31–39. [CrossRef]

28. Donadini, G.; Fumi, M.D.; Newby-Clark, I.R. Consumer's preference and sensory profile of bottom fermented red beers of the Italian market. *Food Res. Int.* **2014**, *58*, 69–80. [CrossRef]

29. Donadini, G.; Porretta, S. Uncovering patterns of consumers' interest for beer: A case study with craft beers. *Food Res. Int.* **2017**, *91*, 183–198. [CrossRef] [PubMed]

30. Stevens, J.F.; Page, J.E. Xanthohumol and related prenylflavonoids from hops and beer: To your good health! *Phytochemistry* **2004**, *65*, 1317–1330. [CrossRef]

31. Araújo, F.B.; Silva, P.H.A.; Minim, V.P.R. Perfil sensorial e composição físico-química de cervejas provenientes de dois segmentos do mercado brasileiro. *Food Sci. Technol.* **2003**, *23*, 121–128. [CrossRef]

32. Giacalone, D.; Frost, M.B.; Bredie, W.L.P.; Pineau, B.; Hunter, D.C.; Paisley, A.G.; Beresford, M.K.; Jaeger, S.R. Situational appropriateness of beer is influenced by product familiarity. *Food Qual. Preferences* **2015**, *39*, 16–27. [CrossRef]

33. Garcia-Burgos, D.; Zamora, M.C. Exploring the hedonic and incentive properties in preferences for bitter foods via self-reports, facial expressions and instrumental behaviours. *Food Qual. Preferences* **2015**, *39*, 73–81. [CrossRef]

Article

A Comparative Study of Dry and Wet Milling of Barley Malt and Its Influence on Granulometry and Wort Composition

Felipe Pereira de Moura and Thiago Rocha dos Santos Mathias *

Laboratory of Fermentation Technology, Federal Institute of Education, Science and Technology of Rio de Janeiro, Rio de Janeiro 20270-021, Brazil; felipemoura.iq.ufrj@gmail.com
* Correspondence: thiago.mathias@ifrj.edu.br; Tel.: +55-21-2566-7772

Received: 5 June 2018; Accepted: 19 July 2018; Published: 20 July 2018

Abstract: Beer is a fermented drink produced from a wort comprised of barley malt, hops, and water in combination with activity from the yeast strains of the genus *Saccharomyces*. The beverage is consumed around the world and has a global market controlled by several multinational companies. However, in recent years, it has been possible to note an increase in the number of microbreweries and homebrewers, necessitating additional research both to develop and increase competitiveness of this market sector as well as to improve product quality and promote the reduction of production costs. The process of milling barley malt is often not considered relevant to these goals; however, this operation is influential with regard to, for example, mashing yield, the concentration of polyphenols in beer, and the quality of wort clarification. Therefore, this work evaluates the wet (10%, 20%, 30%, 40%, and 50% moisture content) and dry barley malt milling process as well as analyzes particle size distribution and the mean diameter of particles. The milled grains were submitted to a mashing process to evaluate how particle size contributes to the conversion of starch to sugars and the availability of polyphenols on sweet wort. The results indicate the best milling conditions to obtain a good mashing yield while preserving as much malt husk as possible to facilitate wort clarification.

Keywords: barley milling; wet milling; brewing technology; granulometry; beer wort

1. Introduction

Beer is produced by the alcoholic fermentation of wort, which is prepared with barley malt and water, hop addition, and yeast action (*Saccharomyces* genus) [1,2]. The process can be divided into the following steps: malting, milling, mashing, boiling, cooling, fermentation, filtration, carbonation, microbiological stabilization, and packaging.

Milling has an important role in the process both in the professional beverage industry and in homebrew production because, in this step, the barley husk breaks down, exposing the starchy endosperm and the content of the embryo (predominantly enzymatic) [3,4]. If the grains are not properly milled, this will influence the rest of the brewing process, notably the wort composition. The presence of intact grains or large fragments result in non-exposure of internal grain fractions, promoting low conversion of starch to fermentable sugars and, consequently, low final yields of the mashing stage [5]. On the other hand, excessive milling promotes the extraction and solubilization of compounds whose presence causes the increase of undesirable characteristics to wort and beer, e.g., sensory properties such as an excessive bitterness or viscosity. Undesirable compounds include phenolic compounds, which, if present in large quantities, cause problems such as excessive bitterness, color changes, and excessive formation of trub [6,7]. Approximately 70–80% of the total polyphenol content of beer comes from the malt husk; its transfer to the wort is influenced by cereal milling [8]. The preservation of grain husk integrity also plays an important role in the formation of the filter cake in the clarification stage of sweet wort [5]. The more intact the husks, the easier and more efficient the clarification step will be.

The milling step can be conducted in either dry or wet form [9]. Dry milling increases grain comminution, which increases the yield of the brewer wort production; however, dry milling can make wort clarification more difficult [4]. Wet milling occurs in the presence of excess water (up to 45%) and promotes greater grain elasticity. Generally, wet milling is used to obtain advantages such as reduction of energy expenditures, elimination of dust (particulate material), transport facilitation, and the reduction of damage to mills (increased durability) [4,10]. However, wet milling can lead to problems, such as greater adhesion of starchy semolina on husk fraction, decreasing the yield. In the case of malt milling, another advantage of wet operation is that the moisture makes the husk more resistant and flexible, which reduces the probability of breakage [2,11,12].

Dry milling is most frequently used by the brewing industry; however, wet milling has recently become another popular method. Currently, wet milling is used industrially for corn and wheat, but it could also be successfully applied to other cereals, such as sorghum, barley, oats, or rice [13]. Wet milling is very common in Africa and Asia but not, for example, in the United States [14] or Brazil. When using non-malted barley as an adjunct to the brewing process, wet milling is the most indicated due to the toughness of this grain [15].

In this context, the objective of this work was to compare the milling process of malt grains with different moisture content to evaluate parameters such as final grain size, availability of starch for conversion to fermentable sugars, and overall content of phenolic compounds. Accordingly, it will be possible to stipulate the best conditions for the milling of barley grains to optimize the brewing process with regard to the mashing yield and the quality of the sweet wort.

2. Materials and Methods

2.1. Materials

Pilsen malt (Agrária®, Paraná, Brazil) was used because it is the base malt for most beer formulations. The malt grain, provided in dry form by the producer (<5%), was obtained in 5 kg sacks and kept dry and sealed throughout the development of the work. The malt specifications were as follows: extract of fine milling (81.5%), diastatic power (292 WK), and protein (11%).

The water used for grain humidification and mashing was potable water from the state's supply system (CEDAE, Rio de Janeiro, Brazil). It was filtered on activated charcoal to remove the free chlorine ions, which result from its chemical treatment.

2.2. Humidification Limit

The water saturation limit was defined to establish the highest moisture percentage used in grain milling. Water was added to a small amount of malt grains (50 g) and the water saturation limit was stipulated as the concentration in which the water was no longer absorbed by the grain and remained at the Becher bottom. This analysis was strictly visual, and the procedure was executed in ambient conditions (25 °C) with constant manual homogenization.

2.3. Milling and Granulometric Analysis

After determining the maximum water absorbed by the grains (Item 2.2), six samples with 50 g of barley malt were prepared in quadruplicate. One sample was kept dry (sample A) and the others were humidified in consideration of the results of the previous analysis, resulting in five samples with the following relation of water/grain: 10%, 20%, 30%, 40%, and 50% (samples B, C, D, E, and F, respectively).

The dry and wet samples were milled in duplicate in cereal disc mill (Guzzo®, Rio Grande do Sul, Brazil), adjusted to 1.0 mm disk spacing. The wet samples were collected and dried in oven at 40 °C until constant mass to avoid contamination and to allow particle size analysis. A mechanical agitator and sequential sieves (SOLOTEST®, Rio de Janeiro, Brazil), with meshes of 2.00 (sieve 1), 1.84 (2), 1.54 (3), 1.20 (4), 0.86 (5), 0.51 (6), and 0.30 mm (7) were used to perform the granulometric analysis [16] of the grains after milling. Previously, the initial quantities were weighed and the masses retained on each sieve were determined as well as the fraction inferior to 0.30 mm, which allowed for a calculation of their respective fractions and to determine the granulometric profile. The mean diameter was calculated from the Sauter mean diameter equation (Equation (1)).

$$\overline{d_p} = \frac{1}{\sum \frac{x_i}{d_{pi}}} \tag{1}$$

where:

d_p = Sauter mean diameter

X_i = mass fraction in sieve i

d_{pi} = particle diameter in sieve i

These results were expressed in bar graphs with particle size distribution profiles and the mean diameter was present with respective deviation. In addition, a linear correlation graph between mean diameter and moisture content for milling was obtained.

2.4. Extraction of Soluble Material and Mashing Milled Grains

Milled grains from each of the six different moisture conditions (dry, 10%, 20%, 30%, 40%, and 50% moisture) were submitted to water-soluble material analysis by simple extraction. For this analysis, the milled grain was washed with excess water (400 mL) at ambient conditions (25 °C) and underwent simple homogenization for 30 min. These samples are called "soluble fraction".

The remaining milled samples of each moisture condition (dry, 10%, 20%, 30%, 40%, and 50% moisture) was reserved for the mashing procedure, following compilation data from the brewing process literature: addition of water (2:1, water to malt ratio) and conduction of the following mash profile curve: 35 °C/10 min (solubilization step), 45 °C/15 min (proteolytic step), 55 °C/15 min (proteolytic step), 66 °C/40 min (β-amylase step), 72 °C/5 min (α-amylase step), and 78 °C/5 min (inactivation step) [1,3,17,18]. These samples were called "sweet wort".

2.5. Analytical Determinations

Subsequently, all "soluble fraction" and "sweet wort" samples obtained in the previous Item (Item 2.3) were centrifuged at 3500 RPM for 3 min under controlled temperature (4 °C) (FANEM®, model 280R, Rio de Janeiro, Brazil). Soluble fractions and sweet worts produced were analyzed by determination of °Plato (by refractometry) [16,19], total reduction of sugar content (by DNS method) [20] and total phenolic compounds content (by Follin–Ciocalteu method) [20], as described below. The chemical analyzes were performed in triplicate and the standard deviation calculated.

For the DNS method, diluted samples were submitted to reaction with DNS reagent (100 °C/10 min) and the resulting product had its absorbance read at 540 nm (BioChrom®, model Libra S-21, Cambridge, UK). The reduced sugar concentration was calculated using a glucose standard curve. For the Follin–Ciocalteu method, diluted samples were reacted with Follin reagent (0.2 N), and Na_2CO_3 4% (50 °C/20 min) and the resulting product had its absorbance read at 740 nm (BioChrom®, model Libra S-21, Cambridge, UK). The total phenolic compounds content was calculated in gallic acid equivalents using a standard curve.

The variation of the parameters before and after mashing was calculated by simple subtracting of the obtained values and was named Δ (delta).

3. Results and Discussion

3.1. Humidification Limit

With humidification and constant homogenization of the grains in ambient conditions, it was possible to observe water deposit in Becher bottom, i.e., water not absorbed by the grains, from 60% moisture sample (60:100 g, water to grain). Therefore, it was decided to perform the study of malt milling with a dry fraction and fractions at 10%, 20%, 30%, 40%, and 50% moisture.

3.2. Granulometric Analysis

The granulometric distribution profiles of each milling condition, dry, 10%, 20%, 30%, 40%, and 50% moisture (samples A to F, respectively) can be seen in Figure 1 and the mean diameter values calculated by Equation (1) are shown in Table

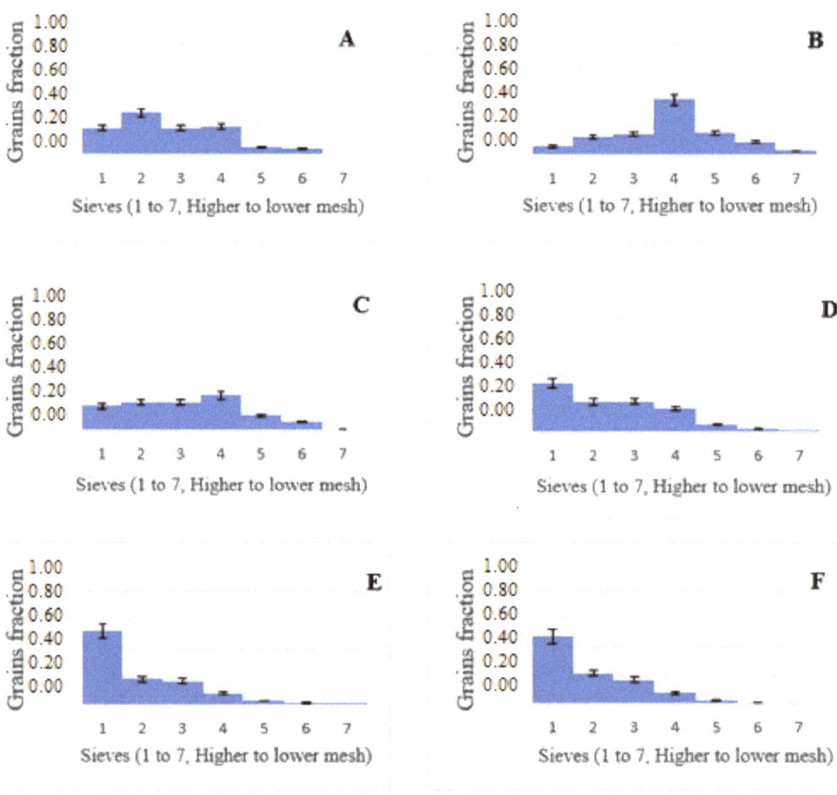

Figure 1. Granulometric distribution profiles of milled grain: dry (**A**); 10% moisture (**B**); 20% moisture (**C**); 30% moisture (**D**); 40% moisture (**E**); and 50% moisture (**F**).

In the dry grain distribution profile (Figure 1A), it was observed that the largest grain fraction (about 30% of the total) remained in the second largest mesh sieve, and the smaller granulometry fractions, sieves 5, 6, and 7, had less than 10% of the total grains. However, the distribution of the grains with 10% moisture (Figure 1B) presented a different granulometric profile from the dry grains (Figure 1A), with a larger fraction (about 40% of the total) in the fourth sieve. It also presented a considerable increase in the mass of particles in the smaller sieves, which determined a reduction in its mean granulometry, as observed by the calculation of the mean diameter (Table 1).

Table 1. Sauter mean diameter (mm) and deviation for each moisture (%) milling condition.

Sample	Moisture	Mean Diameter (mm) *
A	0% (dry)	1.35 ± 0.15
B	10%	1.02 ± 0.20
C	20%	1.24 ± 0.16
D	30%	1.50 ± 0.13
E	40%	1.66 ± 0.12
F	50%	1.66 ± 0.13

* Calculated by Equation (1).

For the samples in the moisture range between 20% and 40% (Figure 1C–F), an increase tendency of accumulated grains in larger sieves was observed. Consequently, the mean diameter of the grains also increased (Table 1). The grain sample with 20% moisture (Figure 1C) showed this tendency, but the largest grain fraction was accumulated in sieve 4, similar to the 10% moisture sample (Figure 1B). For the 30% and 40% moisture (Figure 1D,E), respectively, the largest grain fraction was accumulated in sieve 1.

When the grains with 50% moisture were milled (Figure 1F), the tendency for an increase in grain size was interrupted, and the mean diameter was the same as in the previous sample, with 40% moisture (Table 1). This fact corroborates the hypothesis of water saturation by the grains observed in the previous analysis.

From the calculated mean diameters, the relationship between the mean diameter and the moisture content of the grains was estimated by simple linear regression of the points, excluding the dry and the 50% moisture samples. The results can be seen in Figure 2.

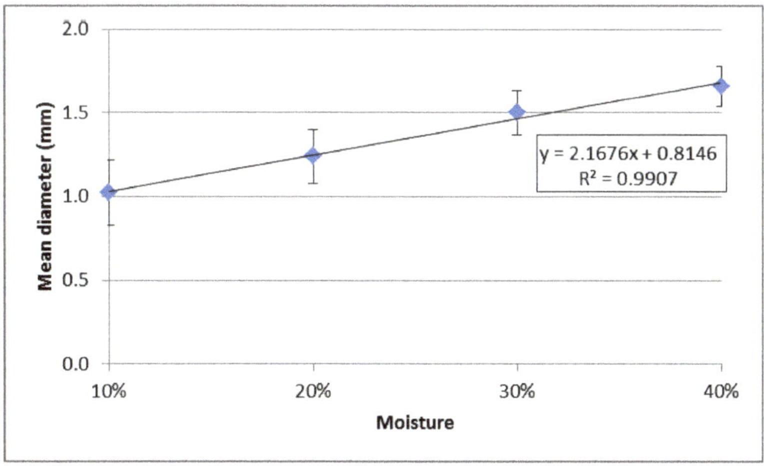

Figure 2. Correlation graph of moisture content (%) for milling with mean grain diameter.

Figure 2 shows a good linear regression ($R^2 = 0.9907$) for the tendency for an increase in mean grain diameter with increasing moisture content between 10% and 40% moisture. The dry sample also had a superior diameter to the samples with 10% and 20% moisture and inferior to the others. The reduction of the granulometry in the 10% moisture sample when compared to dry sample may be attributed to the water content. This amount of water may not have been not able to promote humidification of the whole grain and was only absorbed by the husk. To elucidate what occurs between the dry grain and its humidification in 10% moisture, it is necessary to consult other studies to verify the humidification at other points in this interval. To analyze these two results, it would be interesting to continue the study using fractions of moisture equal to 5%, 15%, and 25%.

In studies by Reinold [5] and Venturini Filho and Nojimoto [12], an increase in the granulometry of the barley malt husk with the adoption of wet milling is described; however, the behavior of the internal fraction of the barley grains is not mentioned. In the literature, the humidification ranges in which an improvement in working conditions occurs are also not determined.

In general, lower values for mean diameter results in an increase in the contact surface of the grains and, consequently, promotes a higher extraction of its components, such as starch (in the case of internal grain fractions) or phenolic compounds (in the case of husk fractions). However, it cannot be disregarded that larger mean diameters, especially for homebrewers, promotes less wort turbidity and, consequently, easier and more efficient clarification.

3.3. Analytical Determinations

The results obtained for both water-soluble wort (before mashing) and sweet wort (after mashing) for refractive indices (°Plato), total reducing sugars, and total phenolic compounds are presented in Table 2. The variation, which refers to the difference of the determinations before and after mashing (mashing yield), was calculated by simple subtraction and is shown in Table 3.

Table 2. Analytical determinations of samples for malted grains with different moisture contents (%) for milling before (soluble fraction) and after mashing (sweet wort).

Sample (% Moisture)	°Plato		TRS [1] (g·L^{-1})		TPC [2] (mg·L^{-1})	
	Soluble Fraction	Sweet Wort	Soluble Fraction	Sweet Wort	Soluble Fraction	Sweet Wort
A (dry)	1.0	9.5	10.16 ± 0.06	89.44 ± 0.00	70.62 ± 2.44	148.99 ± 1.63
B (10%)	3.1	13.9	26.21 ± 1.29	175.84 ± 5.86	131.84 ± 13.08	228.47 ± 0.06
C (20%)	2.8	13.7	25.67 ± 0.06	168.23 ± 3.52	107.42 ± 4.13	215.91 ± 10.29
D (30%)	2.9	12.0	28.14 ± 0.12	146.78 ± 0.59	117.36 ± 4.77	186.74 ± 1.98
E (40%)	3.1	9.1	27.00 ± 1.93	115.10 ± 0.00	119.69 ± 13.95	185.97 ± 10.12
F (50%)	1.9	10.9	17.71 ± 0.47	103.37 ± 4.40	107.75 ± 0.48	188.95 ± 2.87

[1] TRS = Total Reducing Sugars; [2] TPC = Total Phenolic Compounds.

Table 3. Variation (Δ) between parameters before (soluble fraction) and after mashing (sweet wort) calculated by simple subtraction.

Sample (% Moisture)	Δ°Plato	ΔTRS [1] (g·L^{-1})	ΔTPC [2] (mg·L^{-1})
Dry	8.5	79.28 ± 0.05	78.37 ± 0.66
10	10.8	149.63 ± 3.68	96.63 ± 10.63
20	10.9	142.56 ± 2.81	108.49 ± 5.03
30	9.1	118.64 ± 0.31	69.38 ± 2.23
40	6.0	88.1 ± 1.58	66.28 ± 3.12
50	9.0	97.57 ± 3.21	81.20 ± 1.95

[1] TRS = Total Reducing Sugars; [2] TPC = Total Phenolic Compounds.

Table 2 shows that the soluble extract (°Plato) presents a very close initial value for the samples milled with 10%, 20%, 30%, and 40% moisture content, whereas, for the dry sample, this value was significantly decreased, indicating the effect of the wet milling on the exposure of the soluble fractions of the grains. For the produced worts, the highest values of the extract were observed for the samples milled with 10%, 20%, and 30% moisture content.

The results in Table 2 show that the samples with 30% and 40% moisture presents a slightly higher concentration of reducing sugars in the soluble extract than the samples with 10% and 20% moisture. Although not expected because of the higher grain size, this result is close to the expectation related to the determination of °Plato, for which the results of the soluble extract samples with 10%, 20%, 30%, and 40% moisture were very close. After mashing, the highest values of total reducing sugars were observed for the samples milled with 10% and 20% moisture, as these being the smallest particle size, and were according to expected result. The high total values of reducing sugars are very favorable to the process because they guarantee more substrate for the yeast activity and, in turn, allow for the addition of more water in the process, which can maintain an ideal concentration of sugars, leading to a higher overall production of beer with the same initial malt mass. Again, these results indicate that the humidification of the malt provided greater exposure of the grain fractions, allowing the enzyme to be more efficient with the hydrolysis of macromolecules and the liberation of fermentable sugars in the wort.

The results of Table 3 indicate that the higher starch conversion (mashing yield), which can be directly related to the presence of sugars and to soluble solids concentration, is calculated for the samples of 10% and 20% moisture content, which were the samples of smaller particle size. As such, larger deltas of conversion for reducing sugar content of these samples were also calculated. This result was expected because the high contact surface leaves the starchy fraction more available for conversion to sugars via enzyme action. This result indicates that wet milling, using up to 30% moisture, increased the extraction of soluble solids of the grains during mashing.

Szwajgier [21] evaluated the milling (in roller mills) and mashing on an industrial scale of dry and wet malts (12–15%) and observed that there was no difference in the final concentration of sugars (glucose and maltose). However, it was verified that the extraction of these compounds from the grain occurred more rapidly, which may have resulted in a reduction of the mashing duration. Haros and Suarez [22] evaluated corn milling and observed a greater chemical recovery of the starch in wet grains.

Unlike other parameters, in phenols analysis, it is not generally interesting to compare the results with granulometry, given that the contribution of phenols coming from the malt is exclusively from husks, as observed in Broderick et al. [23]. Therefore, a higher concentration of phenols was expected in the samples in which the husks were more comminuted, with no relation to the conservation status of the starchy fraction. Unlike total reducing sugars, the concentration of phenols during mashing is not increased by enzymatic conversion. The increase in the concentration on the stage occurs purely by extraction due to the prolonged contact of the husks with the water and elevation of the temperature.

The 'delta' of extraction exposed in Table 3 suggests that the greatest variation for phenols with mashing occurred in the 10% and 20% moisture samples. The highest result in variations of the sample of 50% moisture as well as in the other parameters analyzed in the present study; again, this result was not expected, as it was necessary to study the behavior of barley malt grains against samples with higher humidification than their moisture saturation. Szwajgier [21] observed that, for total phenolic compounds content, there was a prevention of its extraction from malt husk due to the humidification of the grains.

In conclusion, faster sugar conversion during mashing occurs with wet milling, as confirmed by Szwajgier [21]. Concurrently, greater chemical recovery of the starch occurs for wet barley malts, as confirmed by Haros and Suarez [22] with regard to corn grains milling. To optimize the mashing conditions with an increase in starch conversion to reducing sugars, wet malt milling with 10 or 20% moisture is recommended.

Author Contributions: The experiments were carried out by F.P.d.M. and were supervised by T.R.d.S.M.

Funding: This research received no external funding.

Conflicts of Interest: The authors declare no conflict of interest.

References

1.   Briggs, D.E.; Boulton, C.A.; Brookes, P.A.; Stevens, R. *Brewing Science and Practice*; Woodhead Publishing: Cambridge, UK; CRC Press LLC: Boca Raton, FL, USA, 2004; p. 881, ISBN 0-8493-2547-1.

2. Lewis, M.J.; Young, T.W. *Brewing*, 2nd ed.; Aspen Publishers Inc.: New York, NY, USA, 2001; p. 398, ISBN 978-0306472749.

3. Kunze, W. *Technology Brewing and Malting*, 2nd ed.; VLB: Berlin, Germany, 1999; p. 726, ISBN 978-3-921690-77-2.

4. Mallet, J. *Malt: A Practical Guide from Field to Brewhouse*; Brewers Publications: Boulder, CO, USA, 2014; p. 300, ISBN 193846916X.

5. Reinold, M.R. *Manual Prático de Cervejaria*, 1st ed.; Aden: São Paulo, Brasil, 1997; p. 149, ISBN 85-86395-01-3.

6. Siebert, KJ. Haze Formation in Beverages. *LWT-Food Sci. Technol.* **2006**, *39*, 987–994. [CrossRef]

7. Markovi, R.S.; Grujic, O.S.; Pejin, J.D. Conventional and alternative principles for stabilization of protein and polyphenol fractions in beer. *Acta Period. Technol.* **2003**, *34*, 3–12. [CrossRef]

8. Siqueira, P.B.; Bolini, H.M.A.; Macedo, G.A. O processo de fabricação de cerveja e seus efeitos na presença de polifenóis. *Alim. Nutr. Araraquara* **2008**, *19*, 491–498.

9. Venturi, W.G. *Cerveja. Tecnologia de Bebidas*, 1st ed.; Edgard Blücher: São Paulo, Brazil, 2005; p. 412, ISBN 9788521204930.

10. Cassola, M.S.; Moraes, S.L.; Albertin, E. Desgaste na mineração: O caso dos corpos moedores. *Rev. Esc. Minas* **2006**, *59*, 173–178. [CrossRef]

11. Hough, J.S. *Biotecnología de la Cerveza y de la Malta*; Editora Acribia: Zaragoza, Espana, 1990; p. 208, ISBN 9788420006819.

12. Venturini-Filho, W.G.; Nojimoto, T. Aproveitamento da água de umidificação de malte da moagem úmida como matéria prima na fabricação de cerveja. *Food Sci. Technol.* **1999**, *19*, 174–178. [CrossRef]

13. Wronkowska, M. Wet-milling of cereals. *J. Food Process. Preserv.* **2016**, *40*, 572–580. [CrossRef]

14. Goldammer, T. *The Brewer's Handbook, a Complete Book to Brewing Beer*; Apex Publishers: Herndon, VA, USA, 2008; p. 496, ISBN (13) 978-0-9675212-3-3.

15. Bogdan, P.; Kordialik-Bogacka, E. Alternatives to malt in brewing. *Trends Food Sci. Technol.* **2017**, *65*, 1–9. [CrossRef]

16. American Society of Brewing Chemists (ASBC). *Methods of Analysis*; ASBC: St. Paul, MN, USA, 1976.

17. BARNES, Z.C. Brewing Process Control. In *Handbook of Brewing*; Priest, F.G., Stewart, G.G., Eds.; CRC Press and Taylor & Francis Group: Boca Raton, FL, USA, 2006; pp. 447–486.

18. Bamforth, C.W. *Beer: Tap into the Art and Science of Brewing*, 2nd ed.; Oxford University Press: Oxford, UK, 2003; p. 233, ISBN 978-0195305425.

19. European Brewery Convention (EBC). *Analytica-EBC*, 7th ed.; Fachverlag Hans Carl: Nuremberg, Germany, 2008.

20. AOAC International. *AOAC Official Methods of Analysis*; AOAC International: Rockville, MD, USA, 1990.

21. Szwajgier, D. Dry and Wet Milling of Malt. A Preliminary Study Comparing Fermentable Sugar, Total Protein, Total Phenolics and the Ferulic Acid Content in Non-Hopped Worts. *J. Inst. Brew.* **2011**, *117*, 569–577. [CrossRef]

22. Haros, M.; Suarez, C. Effect of drying, initial moisture and variety in corn wet milling. *J. Food Eng.* **1997**, *34*, 473–481. [CrossRef]

23. Broderick, H.M.; Canales, A.M.; Coors, J.H. *El Cervecero en la Practica: Un Manual para la Industria Cervecera*, 2nd ed.; Associacón de Maestros Cerveceros de las Américas: Provincia Panama, Panama, 1949; p. 230, ISBN 9780971825505.

Review

Computer Vision Method in Beer Quality Evaluation—A Review

Jasmina Lukinac [1], Kristina Mastanjević [1,*], Krešimir Mastanjević [1], Gjore Nakov [2] and Marko Jukić [1]

[1] Faculty of Food Technology Osijek, Josip Juraj Strossmayer University of Osijek, Franje Kuhača 20, 31000 Osijek, Croatia; ptfosptfos2@gmail.com (J.L.); kmastanj@gmail.com (K.M.); ptfosptfos@gmail.com (M.J.)

[2] Department of Biotechnology and Food Technology, University of Ruse Angel Kanchev, Branch Razgrad, Aprilsko vastanie Blvd. 47, Razgrad 7200, Bulgaria; gore_nakov@hotmail.com

* Correspondence: kristinahabschied@gmail.com; Tel.: +385-3122-4356

Received: 8 April 2019; Accepted: 16 May 2019; Published: 1 June 2019

Abstract: Beers are differentiated mainly according to their visual appearance and their fermentation process. The main quality characteristics of beer are appearance, aroma, flavor, and mouthfeel. Important visual attributes of beer are foam appearance (volume and persistence), as well as the color and clarity. To replace manual inspection, automatic, objective, rapid and repeatable external quality inspection systems, such as computer vision, are becoming very important and necessary. Computer vision is a non-contact optical technique, suitable for the non-destructive evaluation of the food product quality. Currently, the main application of computer vision occurs in automated inspection and measurement, allowing manufacturers to keep control of product quality. This paper presents an overview of the applications and the latest achievements of the computer vision methods in determining the external quality attributes of beer.

Keywords: beer; computer vision; image analysis; quality

---

1. Introduction

Beer is one of the oldest (low) alcoholic beverages in the world and the third most widely-consumed drink (after water and tea). Today, besides industrial commercial beers, craft beer has gained great popularity among consumers [1]. Beer is produced in a brewing process from malt, water, brewer's yeast, and hops. The basic quality characteristics of beer are divided into visual and sensory attributes. Appearance is the first attribute that consumer experiences (and evaluates) when he/she gets a glass of beer. Important visual characteristics of beer are foam appearance (volume and persistence), as well as color and clarity of the beer. Perception and acceptability of beer are also determined by other properties e.g., the alcohol content (expressed as Alcohol by volume, ABV or Alcohol by weight, ABW), carbon dioxide content, the presence/absence of off-flavors in bottled beer [2–4], aroma, mouthfeel and bitterness (listed in International Bitterness Units, or IBUs).

The main collections of the standard methods of beer analysis used worldwide are methods of the Institute of Brewing and Distilling (IBD), American Society of Brewing Chemists (ASBC), European Brewery Convention (EBC), and Central European Commission for Brewing Analysis (MEBAK). Beers are differentiated mainly according to their visual appearance (depends on its color and turbidity) and their fermentation process. Regarding the yeast used and the fermentation temperature, beer can be categorized in the three main categories: top-fermented (ale), bottom-fermented (lager), and naturally fermented beer. Beer classification can be done regarding the color during the production process. Beer color is determined by the malt color that develops during the Maillard reactions and caramelization in the malting and roasting process [5]. Therefore, it is possible to distinguish bright, red, dark and black beer color. The aforementioned beer color is actually different shades of yellow, red, brown and black; the most common color is a pale amber produced from pale malts. Commonly used methods for beer classification are flame atomic spectroscopy, linear discriminant analysis [6], nuclear magnetic resonance and multivariate analysis [7,8], and Fourier transform infrared spectroscopy analysis [9]. Trained human inspectors usually perform an inspection of food sensory quality, but this method is unreliable because the results may vary due to subjective evaluations by the inspectors.

Given the unreliability of the human inspectors, there is a need for a precise, objective, and reproducible instrumental method that could imitate human testing methods. Based on this assumption, researches try to develop an instrumental method to evaluate the external and internal appearance of the product non-destructively. Due to a visual inspection, monitoring the external quality attributes is time-consuming, and the computer vision method enables to perform this task automatically. Remembering that a way to improve brewing techniques, rapid feedback from quality control is required. Therefore, the development and implementation of new methods are of great importance in the beer industry. This paper covers an overview of the current achievements in applying the computer vision methods to determine the external quality attributes of beer.

2. Computer Vision and Image Analysis-Basics

Computer vision system (CVS) is a non-destructive technique that includes several different technologies, e.g., optical, mechanical and electromagnetic instrumentation, and digital image processing [10]. CVS is the automatic extraction of information from digital images. CVS is a rapid, objective and effective alternative over the destructive methods [11,12]. Computer vision includes several operations: image capturing, processing and image analysis (Figure 1). After image capture, there is a process of digitization (transformation images into numbers) [13,14].

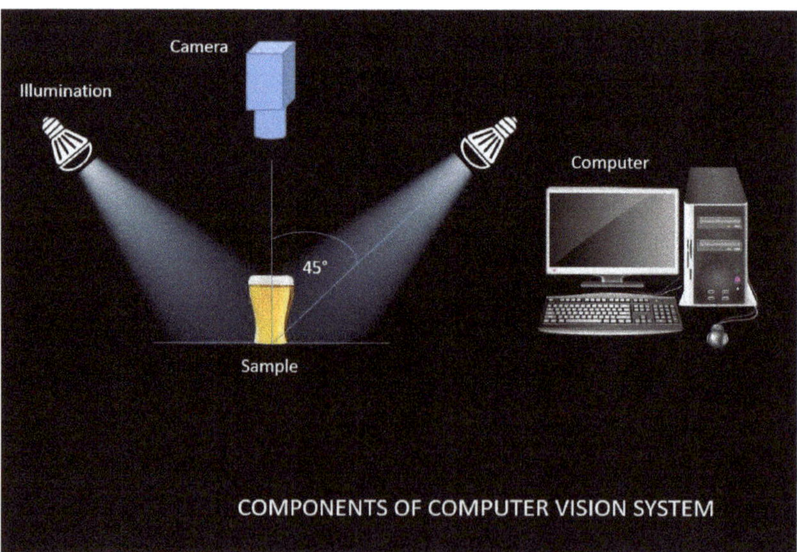

Figure 1. Components of a computer vision system.

CVS can measure the external features of products, recognize objects and extract quantitative information from digital images [15]. Currently, the main application of CVS occurs in the automated inspection providing constant control of product quality. Configuration of CVS is relatively standard, basic components are illumination device (lights), a device for image acquisition (digital camera/scanner), frame grabber (in the case of an analog camera), and computer hardware and software (algorithms for image analyzing and pre-processing) [16,17].

2.1. Illumination

With the camera, one of the most important elements of the CVS system is the illumination. The illumination device is used for object illumination under test. There are many properties of illumination (light angle of incidence, light source color, direct/diffuse light technique), which must be selected in such a way to reduce disturbances (shadows, reflections, noise) or enhance image contrast [14,18,19]. Suitable illumination increases the quality of captured images and improves the accuracy of the analysis. There are several simple rules and good practices that can help select the proper illumination and improve the image quality:

- maximizing the contrast of the features that must be inspected or measured
- minimizing the contrast of the features of no interest
- getting rid of unwanted variations caused by ambient light and differences between items that are non-relevant to the inspection task.

To view an object, it is necessary for it to be illuminated by a light source. The color perception of an object depends on the color of the light source and the color of the surface. Considering the color, we distinguish these light colors: red, green, blue, white with particular color temperature, infrared, and ultraviolet. Light sources may be either natural (sunlight) or artificial (incandescent, halogen, fluorescent, compact fluorescent, LED, metal halide, Xenon, High Pressure Sodium, lasers, and infrared lamps) [20–23]. The most suitable illumination is defined by its position, sources type, geometry and color [14,19]. The term "light source" implies any object that emits energy in the visible spectrum (380–750 nm). Sunlight is one example of the light source, but is not suitable for colorimetric characterization because its quality and energy may vary during the day. To provide reliable colorimetric characterization CIE establishes illuminants upon binding standards (relative energy versus wavelength).

All illuminants are compared against the color temperature, which is described as the spectral energy distribution of the ideal blackbody radiator that radiates light with a specific color at defined temperatures in kelvin (K). Color temperature for an incandescent lamp (~100 W) is 2800 K, for halogen lamp 3000–3200 K, for the fluorescent lamp (cold white) 4000 K, for Xenon and metal halide lamp 4500–5000 K, for warm white LED lamp is 2700–3000 K, and cold white LED lamp 5000–5500 K [20,24].

According to the CIE, we distinguish several illuminants like:

- daylight illuminants lamps (illuminants that represent daylight conditions): C lamp, D50 lamp and D65 lamp
- incandescent/tungsten lamps: A lamp
- fluorescent lamps: F2 lamp (cool white fluorescent), F7 lamp, and F11 lamp
- special light sources.

Several parameters should be considered when deciding on a selection of illuminant source: characteristic of the object surface (whether the object is absorptive, transmissive or reflective), the object geometry (curved of flat), and object/background contrast (translucent objects are usually illuminated with backlight). When an object is illuminated, the incident light is partly (Figure 2):

- reflected and/or
- transmitted and/or
- absorbed and/or
- strayed.

Type of the effect that will occur, depends on the object surface (opaque, semi-transparent or translucent, glossy or matt). Because of that, the position of a light source is very important. When the object surface is glossy, the light is reflected from the object surface at an angle of incidence. If surface is opaque, semi-transparent or translucent, the light is transmitted (penetrates through the object) or it can be polarized or diffracted at the object surface. Usually one of these effects never appears alone, but it is always a combination of several effects.

Due to a different object geometry, there are the three common lighting geometries: the point lighting for flat objects, the diffuse lighting for challenging reflective 3-D objects, and collimated lighting (laser light) [10,22].

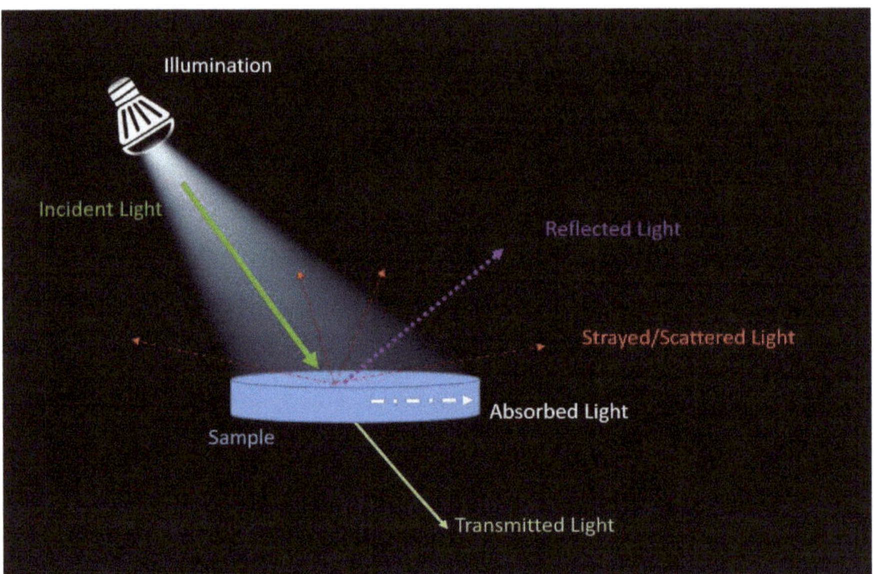

Figure 2. Different types of effects that can occur when the object is illuminated.

Regarding the angle of incidence of the illumination (angle between light source, object and camera), there are several illumination techniques [14]:

- direct incident light (front light)

  - vertical illumination from above,
  - ring illumination,
  - angular illumination;

- incident lighting with a diffuser

  - flat,
  - coaxial,
  - dome-shaped;

- lateral light at angles—at angles from one side or all around;
- shallowly—illumination at a shallow angle from all sides: dark field illumination (usually uses a low angle ring light that is mounted very close to the object where the light rays from the light source are not reflected into the camera lens, but only a proportion of light that is scattered by an uneven surface);
- backlighting—transmitted light from the opposite side of the object (backlit image);
- collimated lighting—laser light (Figures 3–8).

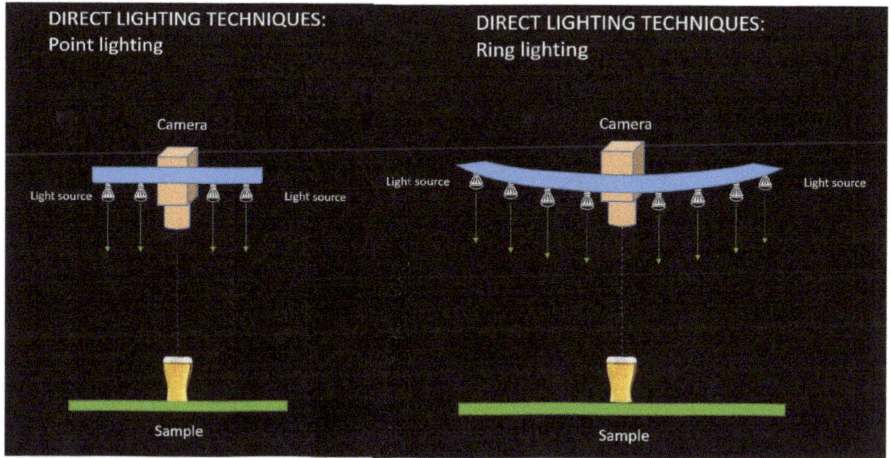

Figure 3. Direct lighting techniques—point and ring lighting.

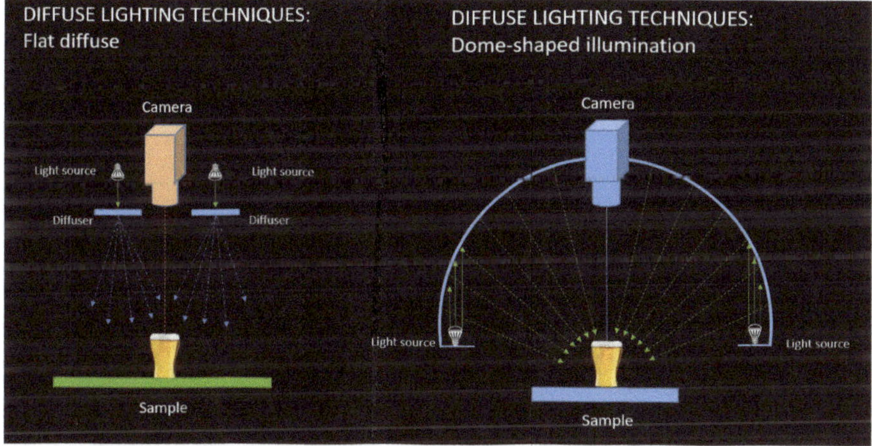

Figure 4. Diffuse lighting techniques—flat diffuse and dome shaped illumination.

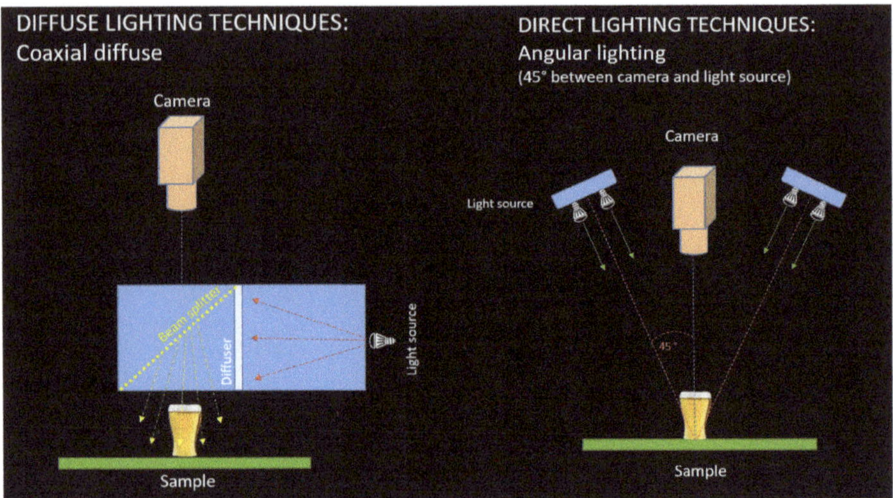

Figure 5. Coaxial lighting techniques with diffuser and angular lighting techniques.

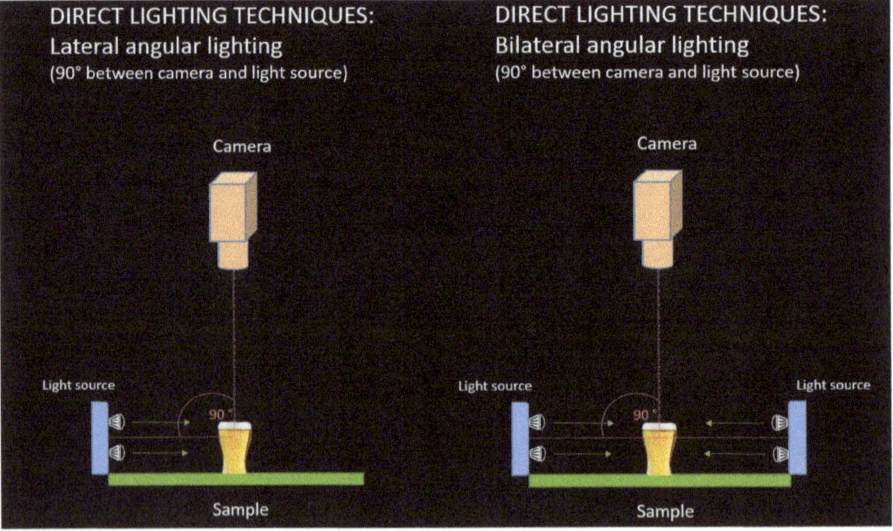

Figure 6. Lateral lighting techniques.

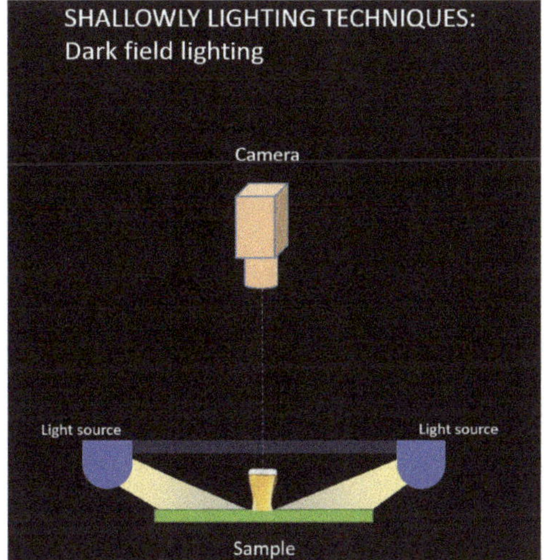

Figure 7. Shallowly lighting techniques—Dark field lighting.

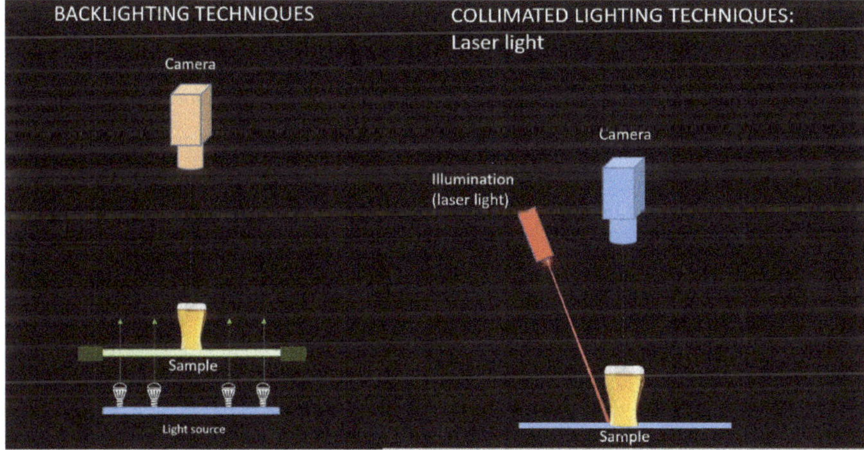

Figure 8. Backlighting and collimated lighting techniques.

2.2. Image Acquisition Devices

Image acquisition encompasses images capturing, with digital camera or scanner. There are three basic elements in image capturing: energy, the lens (optical system) and the sensor. Electronic capturing of the image with camera and frame grabber is the first step in the processing of the digital image. Photons are converted to electrical signals with a camera, and these signals are digitalized with the frame grabber [18]. The frame grabber is required for the transformation of analog to digital signal when analog cameras are used [25]. Usually, only three channels are used by camera (i.e., R, G, B), but when higher accuracy and extended information is required, more channels can be used.

The basic parameters (Figure 9) of an imaging system are:

- Field of View (FOV)—visible object area photographed;
- Working Distance (WD)—the distance between the lens front and the inspected object;

- Resolution—number of pixels or number of minimal parts of the object that can be distinguished by digital imaging;
- Depth of Field (DOF)—the maximum depth tenable in acceptable focus;
- Sensor Size—the active area size of the camera sensor;
- Camera.

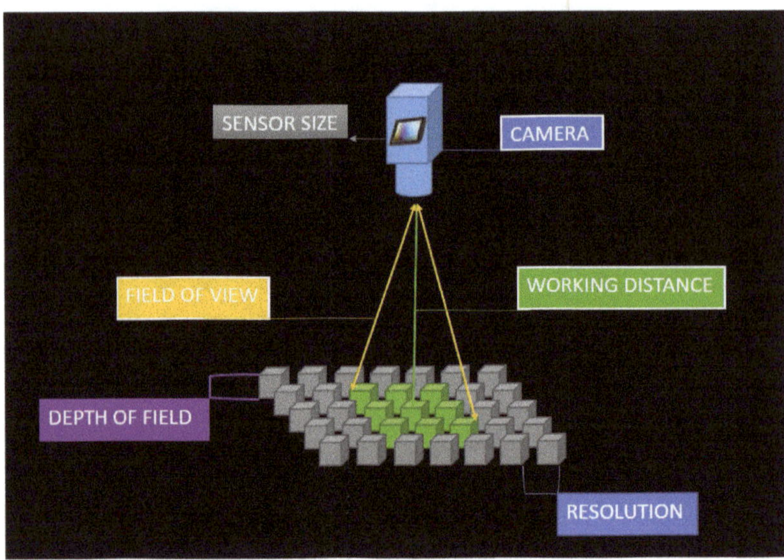

Figure 9. Basic parameters of an Imaging System.

There are a wide range of digital camera capabilities, and consequently, of sizes and prices (Table 1). The most commonly used camera sensors are charged coupled device (CCD) and complementary metal-oxide semiconductor (CMOS) [26]. In a CCD camera, radiation energy (proportional to light exposure) is converted to an electrical signal using of mass photodiodes. These cameras provide good system differentiation and flexibility for various applications (industrial imaging, digital photography, scientific and medical research, etc.). A CMOS camera is more sensitive than a CCD camera, and has a greater dynamic range with very fast signal transferring. They are suitable for industrial use for on-line product inspection, and for all other applications where there is no need for high image quality (wireless video devices, video conferencing, toys, bar-code scanners, etc.) [26].

Scanners are usually used for specialized tasks and specific use (extreme resolution, large numbers of output channels or extreme wavelengths). Qualifying specification parameters for a scanner are bit (color) depth, resolution, and dynamic range. The scanner's bit-depth defines the amount of color information in the image. Accuracy increases with higher depths (4-bit with 16, 8-bit with 256, 24-bit with almost 17 million tones, etc.). Resolution is expressed in dots per inch (dpi) and determines the number of details in the image. Dynamic range is a measure of scanner ability to distinguish image colors and tones. It is similar to image depth and it ranges from perfect white (0.0) to black (4.0) [20].

Table 1. Types of cameras in CVS system.

Cameras Categorization		Application	Features
Cameras Type According to the Image Processing Demands	Network cameras (Internet Protocol camera, IP)	Surveillance	Record a video Day/night modes Special infrared filters Compressed captured photos Simultaneously connected to a large number of users via the Internet
	Industrial cameras (Machine vision) — Area scan cameras / Line scan cameras	Application in different purposes and industries (packaging system management; traffic control system)	Record video or photo Raw photo (no loosing data by compression)
Color	Color cameras Monochrome cameras	intelligent traffic systems	Color image Black and white image
Sensor Types	CMOS sensor	apply in cases where high speed is required	strong value for the performance high frame rates high resolution low power consumption
	CCD sensor	Those sensors are light sensitive, provide image of high quality, and applied in cases where in no need of high speed	high frame rates without deterioration in image quality
Shutter Technique	Global shutter	Capturing fast moving object	allow the light to strike the entire sensor surface all at once
	Rolling shutter	Capturing a stationary object	exposes the image line-by-line
Frame Rate (fps)	Slow sensor	For slow moving or stationary object which require low frame rates (medicine—microscopic inspection)	Small frame rate
	Quick sensor	For fast moving application which require high frame rates (inspection of printed images or labels)	High frame rate
Resolution	Camera with high resolution	For capturing precision image with large number of details	
	Camera with low resolution	For capturing image where details are not the main focus (moving object)	

2.3. Hardware and Supplied Software

Furthermore, important components of CSV are computer hardware and software, providing storage space for images and computing capacity with specific software applications. Visualization of images and results of the image analysis process, are enhanced by high-resolution monitors [16,27–29].

3. Digital Image Analysis

Image capturing (acquisition), processing and analysis are operations included in the computer vision system (Figure 10). The role of image processing is to obtain an image with enhanced quality and to decrease various defects such as noise, distortion, improper lighting and focus, and errors caused by camera movement. Image analysis separates the region of interest (ROI) from the background is then used to obtain quantitative information about the analyzed object. There are several steps in image processing and analysis. The first step is image acquisition and pre-processing of the image to improve its quality (correction of contrast, blur, distortions etc.). Second processing step includes image segmentation (thresholding), representation and description [28]. In the last step, different statistical tools are used to recognize and interpret information obtained from the object image [30].

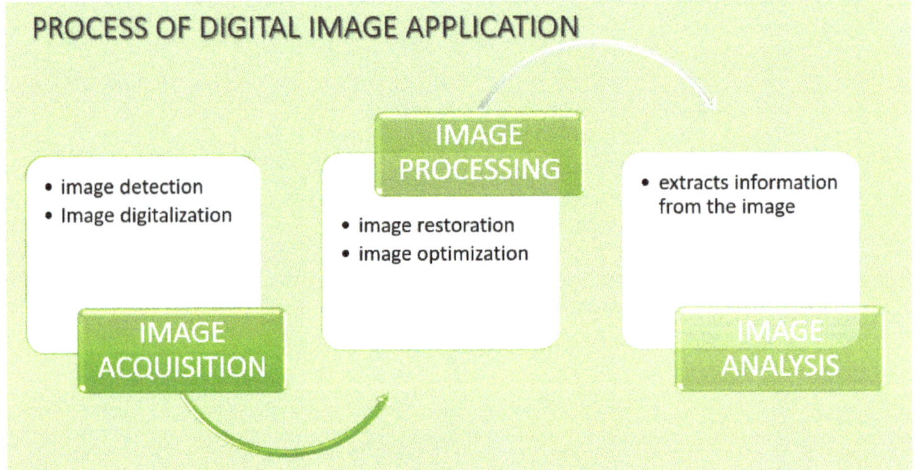

Figure 10. Steps in digital image analysis.

4. Perception and Measuring of Beer Color

Each beer style has a range of acceptable colors. The actual color of each beer is nothing more than gradations of a brown tone, which decreases in concentration through red, copper, and amber colors, through to golden yellow and light yellow. The brewer's ability to predict and control beer color, as one of the three visual attributes which influence beer appearance, is very important. Consumers have a habit associating the beer color with the flavor, and according to the color, to determine which type it belongs, to lager, ale or stout [31]. Consumers expect consistently high-quality standards of foods and beverages, and color loss will be perceived as a sign of quality reduction. They have high requirements in terms of external product quality (appearance, shape, color), so the food industry has the task of providing efficient systems for continuous monitoring of the product's color during the production and storage [12].

The color of the beer comes from malted barley, and wort production. Substances responsible for beer color (melanoidin, polyphenols, trace metals like copper and iron, riboflavin, caramel) are formed during several different chemical reactions [32–34]. Melanoidins are water soluble pigments whose color changes from initially yellow to dark brown [35]. During the kilning process, when the temperature is greater than 95 °C, a Maillard reaction takes place and produces color and aroma components from sugars and amino acids. Caramelization reactions take place at temperatures above 120 °C, and are affected by pH and the type of sugar [36]. At the temperature above 200 °C, pyrolysis reaction occurs and black pigments are produced in the brewing process.

Through history, numerous techniques of color determination of malt and beer have been developed. Grading techniques are based on a comparison of the color standards with the product color. The color is expressed using several scales according to Lovibond (°L), European Brewery Convention color scheme (EBC) [37] or Standard Reference Method (SRM) (Figure 11). A scale for the beer color expression is determined by the used method of color measurement. The first method used to determine the color of beer was the visual comparison method, where beer color is expressed in degrees Lovibond. J.W. Lovibond developed this method in 1893, which was then adopted by Bishop in 1950. The method is quantitative and based on subjective estimation of beer color by visual comparison of beer samples with colored glass discs references. After improving, the European Brewery convention accepted the visual comparison method and the color of malt, wort and beer are expressed in EBC units [31]. In the US, the color of malt is commonly expressed in °L while SRM scale is used for estimation of the beer color. However, visual comparison method ha weaknesses, because it is subjective (based on visual comparison of color sample and reference with the human eye). Because of this disadvantage, new instrumental methods for beer color assessment are developed (spectrophotometric and tristimulus method). That group of methods is independent of the observers, and allow an objective assessment of beer color. In the tristimulus method, light reflected from the sample is separated through the filters into three color channels corresponding to the human eye vision and captured by sensors. Several color systems have been proposed to describe a color as a tristimulus value of individual intensities of red, green and blue [38]. There are many different color systems like Lovibond RYBN color, The Munsell Scale, CIE color system (CIELab, CIELCh, CIEXYZ), RGB color system. The spectrophotometric method uses multiple sensors to measure spectral transmittance or reflectance in a visible light spectrum range (380–740 nm).

Currently, the color of beer and wort is expressed in EBC scale, and North America is an exception where the color is expressed in a SRM scale. The American Society of Brewing Chemists (ASBC) has developed a method for beer color determination by spectrophotometer, called the Standard Reference Method (SRM). The SRM method is based on an absorbance measurement, where the beer color is expressed as a quantity of absorbed light at a wavelength of 430 nm in a 10 mm quartz cuvette against water as a reference [39].

$$A_{430} \cdot D \cdot 12.7 = \text{SRM} \tag{1}$$

where, D = dilution factor of the sample and A_{430} = the light absorbance at 430 nm in a 1 cm cuvette.

Based on spectrophotometry, the European Brewing Convention (EBC) system has developed a similar method in Europe. The simple equation is used to calculate beer color in EBC units:

$$A_{430} \cdot D \cdot 25 = \text{EBC} \tag{2}$$

where, D present dilution factor of the sample, A_{430} present absorbance of the beer at a wavelength of 430 nm in a 1 cm cuvette.

Since EBC and ASBC measurements are based on absorbance at 430 nm, conversion can be made by simply adjusting for the differences in path length and multiplicative factors by the following equation:

$$\text{EBC} = 1.97 \cdot \text{SRM} \tag{3}$$

$$\text{SRM} = \frac{\text{EBC}}{1.97} \tag{4}$$

For both methods, it is important that the beer sample have no turbidity during color measurement. Otherwise, if turbidity is greater than 1 EBC unit, the sample needs to be filtered or centrifuged.

The values of the EBC color scale are approximately 2 times higher than the values of the SRM scale. There are also darker beers whose color is out of the scale and difficult to detect; such samples have to be diluted when measuring [40].

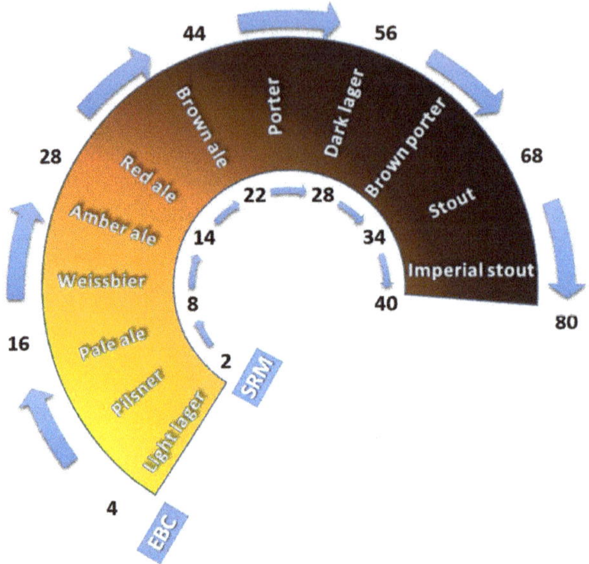

Figure 11. European Brewery Convention (EBC) and Standard Reference Method (SRM) color values of various beer types.

Photometry can be used for color determination of beer and malt [9], and color can be expressed with a tristimulus CIELab color model, were L is the lightness component, a refers to the degree of the red and green, and b to the degree of the blue and yellow color. Lightness (L value) is strongly correlated with the conventional EBC scores [41]. To replace the conventional visual color assessment a novel, computer vision system (CVS) has been developed [24]. CVS enables the non-destructive, contactless, fast, automated, objective, and repeatable color measurement of different food products. It is based on the analysis of each pixel on a captured photo (covering the entire sample), which makes it objective and precise as a method [42,43].

Various methods have been developed for the characterization of different beer types. Silva et al. [44] investigated computer vision method and digital image analysis as an alternative method for color classification of Brazilian pale lager beers. Beer samples are scanned in Petri dishes, and digital image analysis was applied for pattern recognition of beers. After applying digital image analysis, results were presented as color histograms in RGB color space for each brand of Brazilian pale lager beer. Along with conventional analytical methods of color determination, digital image analysis has been demonstrated to be a suitable method for classifying beer of the same type or category. Nikolova et al. [45] reported the use of digital image analysis and spectrophotometer for beer type classification according to their color. For this purpose, several beer samples were chosen to obtain digital images: light beer, dark beer, a beer with lemon and beer with fruits. Cluster analysis was applied to a group of similar beer samples. Beer samples were photographed with a CCD camera in a BMP format. After pre-processing of the image, obtained results were presented as RGB values, and later transformed into the CIELab values. The beer color determined by the image analysis corresponds to the color results measured by the spectrophotometer. Therefore, computer vision and digital image analysis proved capable color-based method in classifying beer.

5. Bubble Size Distribution and Nucleation in Beer

Bubble haze arises in the beer during dispensing beer into the glass (creating an illusion of haze), as a result of a large number of microbubbles which are formed during bubble nucleation. Formed microbubbles do not go immediately to its surface, thus contributing to the effect of the beer bubble haze, and are responsible for the formation of the foam ring on the top of the liquid surface. Bubble haze is an important beer quality parameter and can be preferred by the consumers. For the brewing industry, it is essential to measure beer bubble haze during beer dispense, as one of the product quality indicators. Furthermore, it is very important to determine and control the bubble size in beer haze because their size directly affects the distribution of flavor and taste of beer. Bubbles collect at the top of the glass in a foam called "head". A beer perceived as flat showed little nucleation activity (bubbling) in the glass, regardless of its normal CO_2 content. Also, active bubbling regenerates the head [46–48].

Several methods can measure the frequency of bubble formation, their size and distribution in supersaturated beer liquid. Saxena et al. [49] categorized these methods into three groups:

- Image (photographic) analysis methods—analysis of the captured images of bubbles;
- Optical probe methods—analysis of the bubble penetration length in the area of intensive bubbles migration;
- Electrical conductivity (resistivity) probe methods—analysis of the bubble volume with ultrasound/isokinetic sampling probes.

A digital balance can be used for the estimation of gas loss (measuring the mass loss of the beer glass) over time, to quantify the degree of nucleation activity. The nucleation activity also can be performed visually, observing the reduction in the number of bubbles occurring in beer glass over time. This method is one of the first to be reported for observing the nucleation activity in beer. This method has many faults, is long-lasting, requires a lot of time and is not precise and objective [50]. Lubetkin and Blackwell [51] developed a method for quantification of the nucleation activity using sound data recorded with a microphone and computer at the surface of the liquid. After recording, the sound data were analyzed and the frequency of the bubble rupturing is determined. In this method, the nucleation activity is observed by measuring the sound of rupturing bubbles. This method also has limitations in the application since the beer is the system with foam-forming nature. Image analysis method can also be used for measuring the bubble rate formation in beer. This method is based on image analysis of the photographed bubbles in beer glass during the time. Liger-Belair et al. [52] reported the use of an image analysis method for investigation of the bubble formation in Champagne. They used stroboscopic light and digital camera for image capturing. In this study, the authors reported that the image analysis method could be used to measure numerous parameters of bubble formation: bubble formation frequency, growth rates, and bubbles rising velocities in the liquid.

Hepworth et al. [53–55] made several studies of using computer vision method and image analysis to monitor bubble haze in beer and to investigate the influence of numerous process parameters on bubble haze formation and distribution in beer glass during beer dispense. Hepworth et al. [55] tried to explain the phenomenon of bubble haze in beer more closely in the general aim to predict their presence or absence, and to understand how process parameters affect the bubble haze. They examined the influence of changing the dissolved nitrogen and carbon dioxide content on the stability of the bubble haze in beer. Furthermore, they observed the influence of beer flow rate from the tap and applying a sparkler (a thin plastic disk) in the tap on the stability of the bubble haze. Stability of beer bubble haze was monitored by measuring surge time (the period of the haze disappearing), haze velocity (velocity increase of bubbles), and bubble size distribution (diameter and number of bubbles) in beer. Experiments were carried out using the high-speed video camera and a CCD camera for image acquisition, followed by the image analysis method. Experimental results indicate a positive correlation between the surge time and beer flow rate, sparkler application and supplied nitrogen content. Besides, the size and number of bubbles are crucial factors for the prediction of bubble haze lifetime, and they can be affected by varying the process condition [55].

Hepworth et al. [54] also developed another experimental procedure for monitoring the bubble nucleation (measuring bubble production rate and bubble size) during beer dispense. In this study, they investigated the influence of liquid motion, gas composition, and beer flow rate on the bubble nucleation in beer using computer vision and mathematical modeling. The image analysis was performed on images captured with a CCD camera to determine the bubble diameters and bubble distribution. They adopted the classical mathematical model for predicting bubble nucleation by including liquid motion as a new variable. According to the result, the proposed mathematical model can be suitable for predicting the bubble production rate and size, considering liquid motion. When the flow rate increases, the bubble nucleation rate in beer also increases. The nitrogen content in the gas headspace has an opposite effect on the bubble nucleation (greater nitrogen content, the lower nucleation rate). Hepworth et al. [53] developed and adapted the image analysis method for measuring bubble distribution in beer (measuring bubble production rate and size). For this purpose, they captured images of bubbles in beer, and after image capturing, they applied an image processing to extract valuable data from the captured image like bubble size (diameter) and bubble velocities (distribution and position of bubbles in the observed tank during the time) (Figure 12). In order to ignore the influence of errors and limitations during image capturing (distortion from the curved glass, the distance between the camera lens and glass, focusing problems), they constructed flat-bottomed beer tank with the camera in a very close position. Thus, distortions were avoided due to this positioning method. The proposed method has shown as a good tool for predicting the bubble nucleation rate, bubble growth, and motion under various experimental conditions. The great advantage of this technique is that is designed to be simple to use and relatively portable.

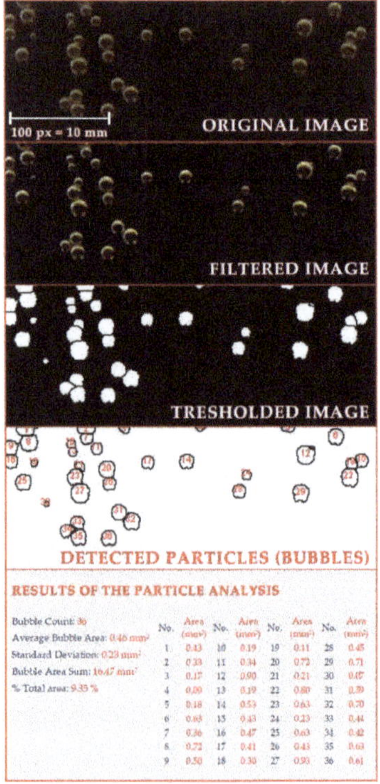

No.	Area (mm²)	No.	Area (mm²)	No.	Area (mm²)	No.	Area (mm²)
1	0.43	10	0.19	19	0.11	28	0.45
2	0.33	11	0.34	20	0.72	29	0.71
3	0.17	12	0.90	21	0.21	30	0.07
4	0.09	13	0.19	22	0.80	31	0.39
5	0.18	14	0.53	23	0.63	32	0.70
6	0.83	15	0.43	24	0.23	33	0.44
7	0.36	16	0.47	25	0.63	34	0.42
8	0.72	17	0.41	26	0.43	35	0.63
9	0.50	18	0.30	27	0.30	36	0.61

Figure 12. Bubble size determination in beer using digital image analysis method.

Zabulis et al. [56] reported the method for bubbles detection and their size determination in dense dispersions with digital image technique. In this study, they developed software for bubble detection, which is easy to use without any need for additional intervention during application. The method is objective and effective in determination of the bubble distribution, and bubble size in beer. This method provides a good tool to control and monitor foam decay through measurement of bubble size distribution. The method is based on monitoring of visual appearance (appearance-based approach) of beer samples with a digital camera in a single image. The proposed approach utilizes templates to increase robustness and an image scale-space to detect bubbles independently of their size. Furthermore, algorithmic optimizations for the proposed approach that target the reduction of computational complexity and user-intervention are proposed and compiled into a software application.

6. Foam Stability (Head Retention)

Beer external quality is characterized by foam stability and visual appearance. Therefore, for beer having a long-lasting head of foam and strong foam, some say that beer is of good quality and is desirable among consumers Features of a good foam are quantity and stability, cling or lacing (adhesion to the glass), density or creaminess, whiteness, and strength. Methods for foam quality evaluation are based on the different foam physical characteristics. Most methods measure foam collapse or the rise of the liquid/foam surface (drainage). Drainage is the liquid flow from a wet foam fraction to the liquid underneath and, in the first stages of the foam decay (collapse), it is the dominant process contributing to foam collapse. However, rearrangement of the foam bubbles during time is an equally important factor when measuring foam stability. Furthermore, some other phenomena also must be considered when beer foam kinetics is studied, e.g., creaming, which represents the rise of the bubbles to the top of the system [57].

Many research activities have been focused on detecting the main components and reaction mechanisms that influence beer foam quality (both positive and negative effects). Stability of beer foam mainly depends on the presence of the hydrophobic surface-active proteins (polypeptides) of albumin class, primarily protein Z (40 kDa) and lipid transfer protein LTP1 (9.7 kDa). Among the proteins mentioned above, the stability of beer foam is also influenced by an iso-α-acids from hops, metal cations, hordein-derived (poly) peptides, polyphenols and non-starch polysaccharides (β-glucan and arabinoxylan). Contrary to substances that promote foam stability, destabilization substances are lipids, some amino acids and increased ethanol content [47,48,56–61]. The beer foam stability can be improved by using nitrogen gas, the way of pouring beer into a glass (pouring from a certain height which can increase the foam in the glass), or by using the beer dispensing system. In addition, the shape of a beer glass plays an important role in beer foam stability. Beer glasses with a larger diameters at the top have a larger exposed surface area of foam in relation to volume ratio, which is why the stability of the foam is disturbed.

The assessment of beer foam quality depends both on the foam generation method, and on foam collapse/drainage measurement procedure. Techniques to generate beer foam can be divided into artificial (foaming devices) and natural categories (pouring). Standard methods for determination of beer foam stability are based on measurements of the weight or volume of the liquid collapse from the foam, visual assessment of foam volume decrease or conductometric assessment of foam/air interface. Some traditional methods for the foam quality assessment are Rudin head retention, Ross and Clark procedure, NIBEM, Sigma head value (SHV), ASBC sigma method [62]. Those traditional (standard) methods can produce errors due to some experimental conditions, e.g., non-reproducible and irregular way of foam generation. Evaluation of the foam stability is much more commonly determined by measuring the liquid drainage rate from the foam due to an inconsistent foam-air boundary, which is difficult to define precisely. Additionally, the beer temperature of 20 °C is greater than one used when beer is served to the customer; foam stability is directly related to temperature (the foam quality differs at lower and higher temperatures).

In contrast to standard methods, new devices like digital cameras and image analysis software have been developed for the assessment of the beer foam stability [59]. Methods using such devices are automated, low-cost, contactless and non-destructive, and measures the foam collapse over time. The method is based on photographing a glass in which a known volume of beers is poured. This method can be used to monitor two parameters simultaneously: beer foam collapse or beer liquid height increase in the glass over the time. The beer poured into a glass and is photographed over time, photos are processed using appropriate image analysis software, and obtained results describe the decrease of the beer foam height and increase of the beer liquid phase height over time. Constant [63] investigated the beer foam collapsing by photographing the foam collapse during a 5 min period. After applied image analysis method on photographed beer samples, he observed two separated parameters on photography: total beer height (which includes foam, and liquid together), and liquid height separated from the foam (beer beneath the foam). Yasui et al. [64] investigated foam stability (foam collapse over time, FTC) using image analysis method. Total foam stability over the time is measured using images captured with CCD camera, and beer samples are poured into a beer glass directly from a beer bottle. This method is based on simulation of the foam formation in realistic conditions (the consumer pours beer directly from the bottle into the beer glass). After images are captured, they are analyzed by the image analysis software. The results of the image analysis method (captured foam collapse during the time) are foam height, FTC and foam stability. They confirmed that image analysis is suitable as a method to monitor foam stability by measuring FTC values. Further, it has been found that monitoring FTC values may replace the usual visual impression of foam stability.

Wallin et al. [65] reported a comparison of three methods for the assessment of foam stability. To measure foam stability of nine different beer, they use NIBEM (foam collapse assessment), Steinfurth (foam drainage assessment) and image analysis technique (monitoring foam stability during time) (Figure 13). The main difference between those methods is a way that foam is generated. In the NIBEM and Steinfurth method, foam is generated artificially (under pressure); meanwhile, for the purpose of the image analysis, foam is generated in a natural pour. The result showed a significant correlation between three different methods for assessing foam stability and good reproducibility, and better foam stability of ale beers over lagers.

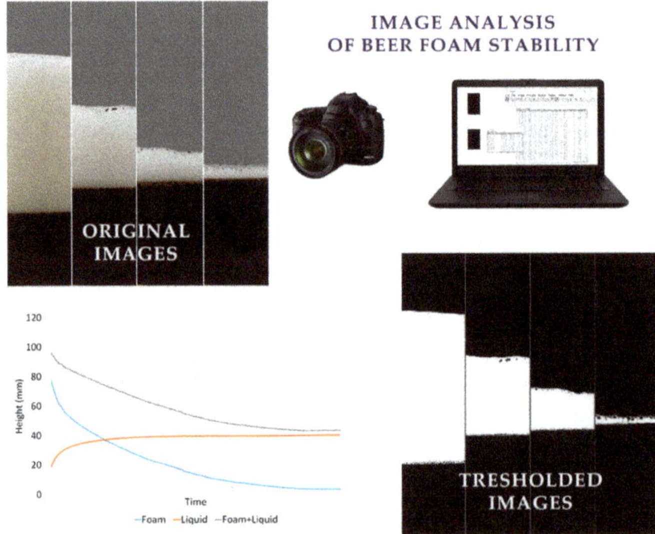

Figure 13. Image analysis of beer foam stability (original images and images with applied threshold segmentation; and graphical parameters presentation of the beer foam stability).

Recently, to analyze some foam properties like foamability, foam stability, and structure, the Dynamic Foam Analyzer (DFA100) has been introduced. This apparatus is based on the principle of the image analysis method. In KRÜSS GmbH study report [66] samples of degassed clear wheat beer, Pils and Kolch were analyzed to determine the influence of the surfactants in the beer independently of the CO_2 content and its impact on the foam behavior. Removal of CO_2 before analyses provides a better insight into the influence of other foam-forming substances (e.g., proteins). The equipment uses a vacuum pump to remove CO_2 from the sample and, after degassing, foaming is conducted in controlled conditions by computer-regulated air pumping through the beer sample. Samples are illuminated with the LED panel and changes in foam height is captured with photodetector of a video camera. During the analysis, the device records the formed foam in a column and simultaneously provides data on the foamability and foam stability. Furthermore, this equipment measures changes in foam structure during the time (size and number of foam bubbles).

Sauerbrei et al. [67] conducted a computer-aided image processing research of redistribution and size of the foam bubbles during the beer foam decay. They used CCD camera, diffuse coaxial illumination to record photographs of the foam collapse, and foam bubble structure over time (5 s interval). The use of the ultrasonic device accomplished foaming. They have successfully implemented image analyses to measure the foam bubble size and area of the liquid fraction in the foam. Immediately after foaming bubbles were very small and uniformly distributed. Over time, coalescence occurs due to rupture of the bubble surface and some of the bubbles merge, forming larger bubbles (Figure 14). Remaining smaller bubbles fill the remaining space between the larger bubbles. The authors have compared the resulting pattern to a known mathematical structure called the Apollonian gasket. They also concluded that beer foam does not follow simple exponential decay but more complex higher order kinetics.

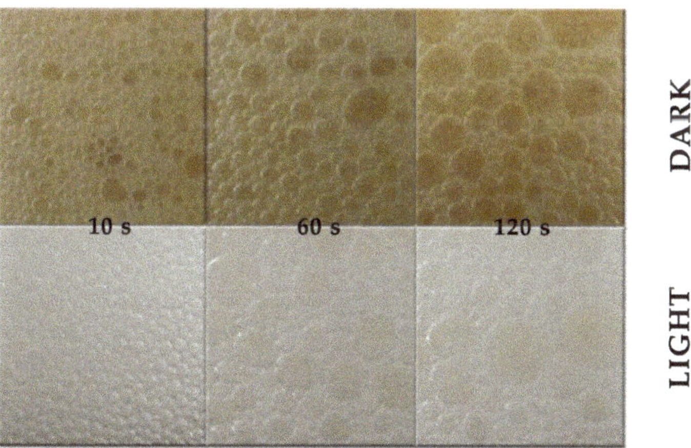

Figure 14. Different bubble size and distribution in the beer foam (10, 60 and 120 s after pouring).

Cimini et al. [68] presented an inexpensive, automated and flexible image analysis method instead of the standard foam head retention method (developed by Rudin [69]). The standard Rudin's method is improved by using three different systems for automatic data collection: tracking software for mouse movement; Accurate Image Analysis system (AIA) and low-cost image analysis (LCIA) system. The main difference between AIA system and LCIA system is that AIA used CCD camera and script written in MATLAB, while LCIA system used a Raspberry Pi single-board computer and a camera module. Authors reported that the LCIA system could be successfully used for analyzing beer foam attributes, such as the number of bubbles in the beer, their distribution, diameter and size.

7. Conclusions

The computer vision system (CVS) represents an efficient and non-destructive inspection technique for evaluating the external attributes of food products. It is an effective alternative to human vision, enabling rapid and objective analyses of food quality properties. Lately, digital image analysis is increasingly used in the study of various beer quality parameters (color, bubble size and distribution, foam stability, etc.). The strongest arguments for deploying CVS in evaluating external quality attributes of beer are the reliability and reproducibility of the obtained results. Although more and more progress is being made on the development of image analysis devices, there is still no available equipment that will simultaneously cover the determination of all quality features, due to the complexity and broad spectrum of beer quality parameters. Despite the fact that CVS is, in any case more, precise than visual methods, algorithms used for image representation and interpretation may be the source of the errors, resulting in the improper extraction and analysis of the obtained data. Therefore, additional efforts must be made in the development of more sophisticated equipment and algorithms that will further enhance measuring precision. One way of improving algorithms is the use of the information collected with the chemometric methods, using a wide set of various statistical tools which are capable of providing useful information by the analysis and modeling large quantities of data.

Author Contributions: Conceptualization, J.L., K.M. (Kristina Mastanjević) and M.J.; Investigation, J.L., K.M. (Kristina Mastanjević), K.M. (Krešimir Mastanjević), G.N. and M.J.; Writing—Original Draft Preparation, J.L., K.M. (Kristina Mastanjević) and M.J.; Writing—Review & Editing, J.L., K.M. (Kristina Mastanjević), K.M. (Krešimir Mastanjević), G.N. and M.J.; Visualization, J.L. and M.J.; Supervision, J.L.

Funding: This research received no external funding.

Conflicts of Interest: The authors declare no conflict of interest. All images shown in this paper are the result of unpublished research of authors of this review paper.

References

1. Mastanjević, K.; Krstanović, V.; Lukinac, J.; Jukić, M.; Lučan, M.; Mastanjević, K. Craft brewing—Is it really about the sensory revolution? *Kvas. Prum.* **2019**, *65*, 13–16. [CrossRef]
2. Gonzalez-Viejo, C.; Fuentes, S.; Li, G.J.; Collmann, R.; Conde, B.; Torrico, D. Development of a robotic pourer constructed with ubiquitous materials, open hardware and sensors to assess beer foam quality using computer vision and pattern recognition algorithms: RoboBEER. *Food Res. Int.* **2016**, *89*, 504–513. [CrossRef] [PubMed]
3. Baert, J.J.; De Clippeleer, J.; Hughes, P.S.; De Cooman, L.; Aerts, G. On the origin of free and bound staling aldehydes in beer. *J. Agric. Food Chem.* **2012**, *60*, 11449–11472. [CrossRef] [PubMed]
4. Cooper, D.J.; Husband, F.A.; Mills, E.N.C.; Wilde, P.J. Role of beer lipid-binding proteins in preventing lipid destabilization of foam. *J. Agric. Food Chem.* **2002**, *50*, 7645–7650. [CrossRef] [PubMed]
5. Ebienfa, P.; Grudanov, V.; Ermakov, A.; Pozdniakov, V. Improving the process of roasting malt with intensive stirring machine. *Ukr. Food J.* **2015**, *4*, 95–108.
6. Bellido-Milla, D.; Moreno-Perez, J.M.; Hernández-Artiga, M.P. Differentiation and classification of beers with flame atomic spectrometry and molecular absorption spectrometry and sample preparation assisted by microwaves. *Spectrochim. Acta Part B* **2000**, *55*, 855–864. [CrossRef]
7. Duarte, I.F.; Barros, A.; Almeida, C.; Spraul, M.; Gil, A.M. Multivariate analysis of NMR and FTIR data as a potential tool for the quality control of beer. *J. Agric. Food Chem.* **2004**, *52*, 1031–1038. [CrossRef]
8. Lachenmeier, D.W.; Frank, W.; Humpfer, E.; Schäfer, H.; Keller, S.; Mörtter, M. Quality control of beer using high-resolution nuclear magnetic resonance spectroscopy and multivariate analysis. *Eur. Food Res. Technol.* **2005**, *220*, 215–221. [CrossRef]
9. Lachenmeier, D.W. Rapid quality control of spirit drinks and beer using multivariate data analysis of Fourier transform infrared spectra. *Food Chem.* **2007**, *101*, 825–832. [CrossRef]
10. Patel, K.K.; Kar, A.; Jha, S.N.; Khan, M.A. Machine vision system: A tool for quality inspection of food and agricultural products. *J. Food Sci. Technol.* **2012**, *49*, 123–141. [CrossRef]
11. Lukinac, J.; Jukić, M.; Mastanjević, K.; Lučan, M. Application of computer vision and image analysis method in cheese-quality evaluation: A review. *Ukr. Food J.* **2018**, *7*, 192–214. [CrossRef]

12. Wu, D.; Sun, D.W. Food colour measurement using computer vision. In *Instrumental Assessment of Food Sensory Quality*, 1st ed.; Kilcast, D., Ed.; Woodhead Publishing Ltd.: Cambridge, UK, 2013; pp. 165–189.

13. Abdullah, M.Z. Image Acquisition System. In *Computer Vision Technology for Food Quality Evaluation*, 2nd ed.; Sun, D.-W., Ed.; Academic Press: San Diego, CA, USA, 2016; pp. 3–43.

14. Brosnan, T.; Sun, D.-W. Improving quality inspection of food products by computer vision—A review. *J. Food Eng.* **2004**, *61*, 3–16. [CrossRef]

15. Sun, D.-W. *Computer Vision Technology for Food Quality Evaluation*, 1st ed.; Sun, D.-W., Ed.; Academic Press: San Diego, CA, USA, 2008.

16. Davies, E.R. *Computer and Machine Vision, Theory, Algorithms, Practicalities*, 4th ed.; Mirmehdi, M., Ed.; Academic Press: Oxford, UK, 2012.

17. Wang, H.-H.; Sun, D.-W. Correlation between cheese meltability determined with a computer vision method and with Arnott and Schreiber. *J. Food Sci.* **2002**, *67*, 745–749. [CrossRef]

18. Moeslund, T.B. Image acquisition. In *Introduction to Video and Image Processing. Building Real Systems and Applications*, 1st ed.; Moeslund, T.B., Ed.; Springer London Ltd.: London, UK, 2012; pp. 7–24.

19. Gunasekaran, S. Computer vision technology for food quality assurance. *Trends Food Sci. Technol.* **1996**, *7*, 245–256. [CrossRef]

20. Batchelor, B.G. *Machine Vision Handbook*, 1st ed.; Batchelor, B.G., Ed.; Springer London Ltd.: London, UK, 2012.

21. Pedreschi, F.; Leon, J.; Mery, D.; Moyano, P. Development of a computer vision system to measure the color of potato chips. *Food Res. Int.* **2006**, *39*, 1092–1098. [CrossRef]

22. Zuech, N. Machine Vision and Lighting. Machine Vision on Line. 2004. Available online: https://www.visiononline. org/vision-resources-details.cfm/vision-resources/Machine-Vision-and-Lighting/content_id/1269 (accessed on 25 January 2019).

23. Abdullah, M.Z.; Abdul-Aziz, S.; Dos-Mohamed, A.M. Quality inspection of bakery products using colour-based machine vision system. *J. Food Qual.* **2000**, *23*, 39–50. [CrossRef]

24. Choudhury, A.K.R. Colour measurement instruments. In *Principles of Colour and Appearance Volume 1: Object Appearance, Colour Perception and Instrumental Measurement*, 1st ed.; Choudhury, A.K.R., Ed.; Woodhead Publishing: Cambridge, UK, 2014; pp. 221–269.

25. Sliwinska, M.; Wisniewska, P.; Dymerski, T.; Namiesnik, J.; Wardenki, W. Food analysis using artificial senses. *J. Agric. Food Chem.* **2014**, *62*, 1423–1448. [CrossRef] [PubMed]

26. Waltham, N. CCD and CMOS sensors. In *Observing Photons in Space. A Guide to Experimental Space Astronomy*, 2nd ed.; Huber, M.C.E., Pauluhn, A., Culhane, J.L., Timothy, J.G., Wilhelm, K., Zehnder, A., Eds.; Springer: New York, NY, USA, 2013; pp. 443–453.

27. Zhang, B.; Huang, W.; Li, J.; Zhao, C.; Fan, S.; Wu, J.; Liu, C. Principles, developments and applications of computer vision for external quality inspection of fruits and vegetables, A review. *Food Res. Int.* **2014**, *62*, 326–343. [CrossRef]

28. Sun, D.W. Inspecting pizza topping percentage and distribution by a computer vision method. *J. Food Eng.* **2000**, *44*, 245–249. [CrossRef]

29. Sun, D.-W. Computer vision-an objective, rapid and non-contact quality evaluation tool for the food industry. *J. Food Eng.* **2004**, *61*, 1–2. [CrossRef]

30. Ying, Y.; Jing, H.; Tao, Y.; Zhang, N. Detecting stem and shape of pears using fourier transformation and an artificial neural network. *Trans. ASAE* **2003**, *46*, 157–162. [CrossRef]

31. Hughes, P.S.; Baxter, E.D. *Beer: Quality, Safety and Nutritional Aspects*; Righelato, R., Ed.; Royal Society of Chemistry: Cambridge, UK, 2001.

32. De Schutter, D.P.; Saison, D.; Delvaux, F.; Derdelinckx, G.; Delvaux, F.R. The chemistry of aging beer. In *Beer in Health and Disease Prevention*, 1st ed.; Preedy, V.R., Ed.; Academic Press Inc.: New York, NY, USA, 2008; pp. 375–388.

33. Lewis, M.J.; Bamforth, C.W. *Essays in Brewing Science*, 1st ed.; Lewis, M.J., Bamforth, C.W., Eds.; Springer: New York, NY, USA, 2006; pp. 20–27.

34. Pozdrik, R.; Roddick, F.A.; Rogers, P.J.; Nguyen, T. Spectrophotometric method for exploring 3-methyl-2-butene-1-thiol (MBT) Formation in Lager. *J. Agric. Food Chem.* **2006**, *5*, 6123–6129. [CrossRef]

35. Nursten, H.E. *The Maillard Reaction, Chemistry, Biochemistry and Implication*, 1st ed.; Baynes, J.W., Ed.; The Royal Society of Chemistry: Cambridge, UK, 2005.

36. Park, C.W.; Kang, K.O.; Kim, W.J. Effects of reaction conditions for improvement of caramelization rate. *Korean J. Food Sci. Technol.* **1998**, *30*, 983–987.

37. Committee of the EBC. *Analytica EBC*; Schweizer Brauerei-Rundschau: Zurich, Switzerland, 1975.

38. Smedley, S.M. Colour determination of beer using tristimulus values. *J. Inst. Brew.* **1992**, *98*, 497–504. [CrossRef]

39. De Lange, A.J. Color. In *Brewing Materials and Processes. A Practical Approach to Beer Excellence*, 1st ed.; Bamforth, C.W., Ed.; Academic Press Elsevier: London, UK, 2016; pp. 199–249.

40. De Lange, A.J. The standard reference method of beer color specification as the basis for a new method of beer color reporting. *J. Am. Soc. Brew. Chem.* **2008**, *66*, 143–150.

41. Davies, N.L. Perception of color and flavor in malt. *MBAA Tech. Q.* **2010**, *474*, 12–16. [CrossRef]

42. Yagiz, Y.; Balaban, M.O.; Kristinsson, H.G.; Welt, B.A.; Marshall, M.R. Comparison of Minolta colorimeter and machine vision system in measuring colour of irradiated Atlantic salmon. *J. Sci. Food Agric.* **2009**, *89*, 728–730. [CrossRef]

43. Leon, K.; Mery, D.; Pedreschi, F.; Leon, J. Color measurement in L*a*b* units from RGB digital images. *Food Res. Int.* **2006**, *39*, 1084–1091. [CrossRef]

44. Silva, T.C.O.; Godinho, M.S.; De Oliveira, A.E. Identification of pale lager beers via image analysis. *Lat. Am. Appl. Res.* **2011**, *41*, 141–145.

45. Nikolova, K.T.; Gabrova, R.; Boyadzhiev, D.; Pisanova, E.S.; Ruseva, J.; Yanakiev, D. Classification of different types of beer according to their colour characteristics. *J. Phys. Conf. Ser.* **2017**, *794*, 1742–6596. [CrossRef]

46. Mastanjević, K.; Krstanović, V.; Lukinac, J.; Jukić, M.; Vulin, Z.; Mastanjević, K. Beer–The Importance of colloidal stability (non-biological haze). *Fermentation* **2018**, *4*, 91. [CrossRef]

47. Bamforth, C.W. The relative significance of physics and chemistry for beer foam excellence: Theory and practice. *J. Inst. Brew.* **2004**, *110*, 259–266. [CrossRef]

48. Bamforth, C.W. The foaming properties of beer. *J. Inst. Brew.* **1985**, *91*, 370–383. [CrossRef]

49. Saxena, S.C.; Patel, D.; Smith, D.N.; Ruether, J.A. An assessment of experimental techniques for the measurement of bubble size in a bubble slurry reactor as applied to indirect coal liquefaction. *Chem. Eng. Commun.* **1988**, *63*, 87–127. [CrossRef]

50. Shafer, N.E.; Zare, R.N. Through a beer glass darkly. *Phys. Today* **1991**, *44*, 48–52. [CrossRef]

51. Lubetkin, S.; Blackwell, B. The nucleation of bubbles in supersaturated solutions. *J. Colloids Interface Sci.* **1988**, *26*, 610–615. [CrossRef]

52. Liger-Belair, G.; Marchal, R.; Robillard, B.; Vignes-Adler, M.; Maujean, A.; Jeandet, P. Study of effervescence in a glass of champagne: Frequencies of bubble formation, growth rates, and velocities of rising bubbles. *Am. J. Enol. Vitic.* **1999**, *50*, 317–323.

53. Hepworth, N.J.; Hammond, J.R.M.; Varley, J. Novel application of computer vision to determine bubble size distributions in beer. *J. Food Eng.* **2004**, *61*, 119–124. [CrossRef]

54. Hepworth, N.J.; Boyd, J.W.R.; Hammond, J.R.M.; Varley, J. Modelling the effect of liquid motion on bubble nucleation during beer dispense. *Chem. Eng. Sci.* **2003**, *58*, 4071–4084. [CrossRef]

55. Hepworth, N.J.; Varley, J.; Hind, A. Characterizing gas bubble dispersions in beer. *Food Bioprod. Process.* **2001**, *79*, 13–20. [CrossRef]

56. Zabulis, X.; Papara, M.; Chatziargyriou, A.; Karapantsios, T.D. Detection of densely dispersed spherical bubbles in digital images based on a template matching technique—Application to wet foams. *Colloids Surf. A Physicochem. Eng. Asp.* **2007**, *309*, 96–106. [CrossRef]

57. Evans, D.E.; Bamforth, C.W. Beer foam: Achieving a suitable head. In *Beer: A Quality Perspective*, 1st ed.; Bamforth, C.W., Ed.; Academic Press: Burlington, NJ, USA, 2009; pp. 1–60.

58. Asano, K.; Hashimoto, N. Isolation and characterization of foaming proteins of beer. *J. Am. Soc. Brew. Chem.* **1980**, *38*, 129–137. [CrossRef]

59. Evans, D.E.; Sheehan, M.C. Do not be fobbed off, the substance of beer foam, a review. *J. Am. Soc. Brew. Chem.* **2002**, *60*, 47–57.

60. Kunimune, T. Foam Enhancing Properties of Hop Bitter Acids and Propylene Glycol Alginate. Master's Thesis, Oregon State University, Corvallis, OR, USA, 14 January 2007.

61. Hao, J.L.Q.; Dong, J.; Yu, J.; Fan, W.; Chen, J. Identification of the major proteins in beer foam by mass spectometry following sodium dodecyl sulphatepolyacrylamide gel electrophoresis. *J. Am. Soc. Brew. Chem.* **2006**, *64*, 166–174.

62. Evans, E.D.; Surrel, A.; Sheehly, M.; Stewart, D.C.; Robinson, L.H. Comparison of foam quality and the influence of hop alpha-acids and proteins using five foam analysis methods. *J. Am. Soc. Brew. Chem.* **2008**, *66*, 1–10.

63. Constant, M. A practical method for characterizing poured beer foam. *J. Am. Soc. Brew. Chem.* **1992**, *50*, 37–47. [CrossRef]

64. Yasui, K.; Yakoi, S.; Shigyo, T.; Tamaki, T.; Shinotsuka, K.A. Customer-orientated approach to the development of a visual and statistical foam analysis. *J. Am. Soc. Brew. Chem.* **1998**, *56*, 152–158.

65. Wallin, C.E.; Di Pietro, M.B.; Schwarz, R.W.; Bamforth, C.W. A comparison of three methods for the assessment of foam stability of beer. *J. Inst. Brew.* **1992**, *116*, 78–80. [CrossRef]

66. Kruss Application Report: Comparison of The Foam Behaviour of Different Types of Beer Independently of CO_2 Content. (Kruss, AR275). Available online: https://www.kruss-scientific.com/fileadmin/user_upload/website/literature/kruss-ar275-en.pdf (accessed on 25 January 2019).

67. Sauerbrei, S.; Haß, E.C.; Plath, P.J. The Apollonian decay of beer foam bubble size distribution and the lattices of young diagrams and their correlated mixing functions. *Discret. Dyn. Nat. Soc.* **2006**, *2006*, 79717. [CrossRef]

68. Cimini, A.; Pallottino, F.; Menesatti, P.; Moresi, M. A low-cost image analysis system to upgrade the rudin beer foam head retention meter. *Food Bioprocess. Technol.* **2016**, *9*, 1587–1597. [CrossRef]

69. Rudin, A.D. Measurement of the foam stability of beers. *J. Inst. Brew.* **1957**, *63*, 506–509. [CrossRef]

 *beverages*

Article

Development of Artificial Neural Network Models to Assess Beer Acceptability Based on Sensory Properties Using a Robotic Pourer: A Comparative Model Approach to Achieve an Artificial Intelligence System

Claudia Gonzalez Viejo *, Damir D. Torrico, Frank R. Dunshea and Sigfredo Fuentes

School of Agriculture and Food, Faculty of Veterinary and Agricultural Sciences, University of Melbourne, Parkville, VIC 3010, Australia; damir.torrico@unimelb.edu.au (D.D.T.); fdunshea@unimelb.edu.au (F.R.D.); sfuentes@unimelb.edu.au (S.F.)
* Correspondence: cgonzalez2@unimelb.edu.au; Tel.: +61-412-055-704

Received: 6 March 2019; Accepted: 9 April 2019; Published: 1 May 2019

Abstract: Artificial neural networks (ANN) have become popular for optimization and prediction of parameters in foods, beverages, agriculture and medicine. For brewing, they have been explored to develop rapid methods to assess product quality and acceptability. Different beers ($N = 17$) were analyzed in triplicates using a robotic pourer, RoboBEER (University of Melbourne, Melbourne, Australia), to assess 15 color and foam-related parameters using computer-vision. Those samples were tested using sensory analysis for acceptability of carbonation mouthfeel, bitterness, flavor and overall liking with 30 consumers using a 9-point hedonic scale. ANN models were developed using 17 different training algorithms with 15 color and foam-related parameters as inputs and liking of four descriptors obtained from consumers as targets. Each algorithm was tested using five, seven and ten neurons and compared to select the best model based on correlation coefficients, slope and performance (mean squared error (MSE). Bayesian Regularization algorithm with seven neurons presented the best correlation ($R = 0.98$) and highest performance (MSE = 0.03) with no overfitting. These models may be used as a cost-effective method for fast-screening of beers during processing to assess acceptability more efficiently. The use of RoboBEER, computer-vision algorithms and ANN will allow the implementation of an artificial intelligence system for the brewing industry to assess its effectiveness.

Keywords: beer acceptability; machine learning; robotics; fast-screening; automation

---

1. Introduction

Machine learning is defined as the computer-based system that is able to learn and find patterns among the data to predict specific outputs [1,2]. There are different types of machine learning from which two main categories are derived: (i) pattern recognition or classification and (ii) fitting or regression [3]. The first is mainly used for decision making as it classifies samples into two or more categories, the most publicized applications can be found in medical diagnosis [4,5], food and beverages to classify into types of brewages [6–8] and level of liking of brewages [8,9], in agriculture for identification of grapevine cultivars [10], and to estimate plant water status [11], among others. Fitting or regression is used to predict specific values of certain variables such as chemical compounds [7,12], sensory descriptors [13], and microbial spoilage [14] among others.

There are different types of regression algorithms, which can be classified within categories such as linear regression, regression trees, support vector machines, Gaussian process, ensemble of trees and artificial neural networks (ANN) [15]. The latter has been widely used due to its non-linearity and ability to find patterns from inputs in a similar way to the functioning of neurons in the human brain. These algorithms are able to learn from data by testing and modifying weights and biases until they find the best correlation [7,16]. Furthermore, it has the advantage that the derived ideal relationship, which links the inputs and outputs, is obtained during the training stage [17,18]. There are several ANN training algorithms that may be used, which can be classified into four main categories: (i) backpropagation with Jacobian derivatives, (ii) backpropagation with gradient derivatives [3], (iii) supervised weight and bias, and (iv) unsupervised weight and bias training functions [19]. In this paper, only the first three categories will be used.

The use of machine learning algorithms, especially ANN, in food and brewages has become more popular in recent years as they aid in the increase in accuracy, time and cost reduction in analytical and sensory methods to assess quality and acceptability of beverages [20]. Specifically, in beer, it has been used in the prediction of chemical compounds using near-infrared spectroscopy [7,21,22], and prediction of the intensity of sensory descriptors [13,23].

This paper aimed to find the best machine learning regression model by comparing 17 different ANN training algorithms to predict the liking of four sensory attributes of beer using 15 color and foam-related parameters measured using a robotic pourer (RoboBEER), and computer vision algorithms [6]. For this purpose, 17 beer samples from different styles and from the three types of fermentation (top, bottom and spontaneous) were analyzed in triplicates to develop the models. The targets considered for the models were obtained conducting a sensory session with 30 consumers which rated the liking of four attributes (carbonation mouthfeel, bitter taste, flavor and overall liking). After comparing the models developed using the 17 training algorithms, the best model was selected on the basis of best performance. The best models found may potentially be used for fast-screening of beer samples in product development and/or at the end of the production line to assess beer acceptability without the need of recruiting consumers, which is more cost-effective and less time-consuming.

2. Materials and Methods

2.1. Beer Samples Description

Triplicates of 17 different beer samples ($N = 51$) from different countries, styles and type of fermentation (Table 1) were used to assess their color and foam-related parameters. However, only one replicate was used to assess consumer acceptability as the replicates were obtained from bottles belonging to the same production batch.

Table 1. List of samples used for the study, indicating their style, country of origin and type of fermentation.

Beer Style	Country of Origin	Type of Fermentation
Kolsch	Australia	Top
Porter	Poland	Top
Steam Ale	Australia	Top
Sparkling Ale	Australia	Top
Blonde Ale	Belgium	Top
Red Ale	USA	Top
American Lager	Mexico	Bottom
American Lager	Mexico	Bottom
Lager	The Netherlands	Bottom
Pilsner	Czech Republic	Bottom
American Lager	USA	Bottom
Pilsner	Czech Republic	Bottom
Lambic Gueuze	Belgium	Spontaneous
Lambic Cassis	Belgium	Spontaneous
Lambic Kriek	Belgium	Spontaneous
Lambic Framboise	Belgium	Spontaneous

2.2. Color and Foam-Related Parameters

Color and foam-related parameters were obtained using a robotic pourer, RoboBEER (University of Melbourne, Melbourne, Australia), to ensure uniform pouring. RoboBEER works with two Lego® servo motors and has three sensors attached that work with Arduino® (Arduino, Ivrea, Italy): (i) temperature, (ii) alcohol and (iii) carbon dioxide (CO_2) gas release and is coupled with an iPhone 5S to record 5 min videos of the pouring (Figure 1). These videos were then analyzed with Matlab® R2018b (Mathworks Inc., Matick, MA, USA) using customized computer vision algorithms. The first algorithm worked in a semi-automatic way, which consisted of standardizing and scaling the glass size by selecting the height and glass rim in the first frame of the video, followed by the manual selection of the foam height every 30 frames for the algorithm to automatically calculate the foam and beer volume. These results were then used to develop the foam volume versus time curve and to calculate the following parameters: (i) maximum volume of foam (MVol), (ii) total lifetime of foam (TLTF), (iii) lifetime of foam (LTF), and (iv) foam drainage (FDrain). Furthermore, a single frame of the video (highest in foam) was processed using other algorithms in Matlab® to assess color in two scales CIELab [(v) L, (vi) a, (vii) b] and RGB [(viii) R, (ix) G, (x) B] as well as bubble size distribution divided in (xi) small (SmB), (xii) medium (MedB), and (xiii) large bubbles (LgB), the latter were analyzed based on the "Hough Transformation" from the middle section of the foam and classifying bubble size based on the diameter measured in pixels. Additionally, the parameters (xiv) alcohol (OH) and (xv) CO_2 gas release from the sensors were obtained. More details about the robotic pourer and computer vision analysis can be found in the paper from Gonzalez Viejo et al. [6]. All data were analyzed using customized codes in Matlab® and a Titan Xp GPU (NVIDIA Corporation, Santa Clara, CA, USA).

Figure 1. *Cont.*

Figure 1. Equipment used to assess beers physical measurements; (**a**) robotic pourer, RoboBEER, which was used to assess the color and foam-related parameters and (**b**) a frame of a video taken to analyze the beer using computer vision algorithms.

2.3. Sensory Session

A double-blind sensory session to assess beer acceptability was conducted with 30 consumers using a 9-point hedonic scale. According to the Power analysis, this sample size of consumers is enough to compare samples in a sensory test $(1-\beta > 0.99)$. The session was conducted in individual booths with uniform lighting located in the sensory laboratory of the Faculty of Veterinary and Agricultural Sciences of The University of Melbourne. Before the sensory session, participants were asked to sign a consent form in accordance with the ethics approval 1545786.2 by the Human Ethics Advisory Group (HEAG) of the Faculty of Veterinary and Agricultural Science at The University of Melbourne. The beer samples were semi-randomized in two blocks of eight and nine samples at refrigeration temperature (4 °C) and participants were provided with crackers and water to cleanse the palate and to allow them to rest between samples to avoid fatigue. The sensory attributes evaluated and used as targets for the model construction consisted of (i) carbonation mouthfeel (MCarb), (ii) bitter taste (TBitt), (iii) flavor, and (iv) overall liking (overall).

2.4. Machine Learning Modelling

Seventeen training algorithms (Table 2) were used to develop artificial neural network models using a customized Matlab® code capable of testing all the algorithms in a loop. The models were developed using as inputs the normalized values (from −1 to 1) of the 15 color and foam-related parameters measured with the RoboBEER: (i) MVol, (ii) TLTF, (iii) LTF, (iv) FDrain, (v) L, (vi) a, (vii) b, (viii) R, (ix) G, (x) B, (xi) SmB, (xii) MedB, (xiii) LgB, (xiv) OH and (xv) CO_2, and the four sensory attributes as targets/outputs: (i) MCarb, (ii) TBitt, (iii) flavor, and (iv) overall.

Table 2. Algorithms used and description of the main function type and abbreviations, which were used to develop the artificial neural network models.

Main Function Type	Algorithm	Abbreviation
Backpropagation with Jacobian derivatives	Levenberg Marquardt	LM
	Bayesian Regularization	BR
Backpropagation with gradient derivatives	Broyden, Fletcher, Goldfarb, and Shanno quasi-Newton	BFGS
	Conjugate gradient with Powell-Beale restarts	PB
	Conjugate gradient with Fletcher-Reeves updates	FR
	Conjugate gradient with Polak-Ribiere updates	PR
	Gradient descent backpropagation	GD
	Gradient descent with adaptive learning rate	GDLR
	Gradient descent with momentum	GDM
	Gradient descent with momentum and adaptive learning rate	GDMLR
	One step secant	OSS
	Resilient backpropagation	RPROP
	Scaled conjugate gradient	SCG
Supervised weight and bias training functions	Batch training with weight and bias learning rate	BLR
	Cyclical order weight and bias	CO
	Random order weight and bias	RO
	Sequential order weight and bias	SO

A neuron trimming exercise (5, 7 and 10 neurons) was performed for each algorithm. Ten was the largest number of neurons tested as using fewer neurons and obtaining good models without overfitting is the best practice. Using a larger number of neurons would most likely lead to overfitting. All models were developed using a random data division considering 70% ($n = 35$) of samples used for training, 15% ($n = 8$) for validation using a mean squared error performance algorithm, and 15% ($n = 8$) for the testing stage with a default derivative function. The models were constructed based on a two-layer feedforward network with a tan-sigmoid function in the hidden layer and a linear transfer function in the output layer (Figure 2).

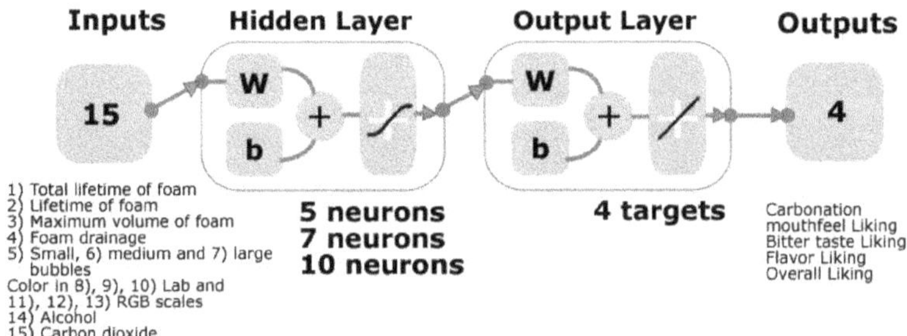

Figure 2. A two-layer feedforward model diagram showing the 15 inputs, number of neurons tested in the hidden layer, and targets/outputs used to create the model.

The statistical analysis to evaluate and compare the accuracy of the models developed consisted of the correlation coefficient (R), determination coefficient (R^2), mean squared error (MSE) to assess performance and slope (b) for each stage (i) training, (ii) validation, (iii) testing, and (iv) overall model as well as the p-value for the overall model. For the three best models, the percentage of outliers using 95% confidence bounds were obtained.

3. Results

Table 3 shows the statistical data of the best and worse models developed from each group of training algorithms. For the backpropagation with Jacobian derivatives algorithm, there was no worse model as those from both algorithms within the group produced two of the best models. Tables S1–S3 in Supplementary Material show the statistical data of the models developed using the 17 training algorithms. Correlations from all models were significant with a *p*-value < 0.0001. It can be observed that the algorithms with the lowest R and R^2 were from the gradient descent backpropagation with five and seven neurons (Table 3; Table S1), the batch training with weight and bias learning rate with seven neurons (Table 3) and the sequential order weight and bias with five neurons (Table S3). On the other hand, the models with the highest R and R^2 were with those developed using seven neurons from both algorithms belonging to the backpropagation with Jacobian derivatives function (LM and BR) and the RPROP with R values consistently over 0.90 for all stages (Table 3). Furthermore, the slope from these three best models was close to unity ($b \sim 1$) for all stages, with the RPROP having the lowest slope values with a $b = 0.90$ for the overall model (Table 3; Figure 3). On the other hand, the three models had low MSE values ($\leq$0.06) for the three stages and overall model. Table 3 also shows the best model from the supervised weight and bias algorithms; however, this still had some signs of overfitting as the validation and testing performances were not as close (MSE = 0.10 and 0.06, respectively) and the R values were lower than the three best models.

Table 3. Statistical results of the best and worse models developed using the algorithms from the three different groups. Numbers in bold represent the models with the highest correlation and determination coefficients from each group of algorithms.

Algorithm	Neurons	Stage	R	R^2	b	MSE
Backpropagation with Jacobian derivatives algorithm						
Levenberg Marquardt	7	Training	**0.96**	**0.92**	**0.94**	**0.02**
		Validation	**0.95**	**0.90**	**1.00**	**0.06**
		Testing	**0.95**	**0.90**	**1.10**	**0.05**
		Overall	**0.95**	**0.90**	**0.98**	**0.03**
Bayesian Regularization	7	Training	**0.99**	**0.98**	**0.97**	**0.01**
		Validation	-	-	-	-
		Testing	**0.97**	**0.94**	**1.1**	**0.03**
		Overall	**0.98**	**0.96**	**1.0**	**0.01**
Backpropagation with gradient derivative algorithms						
Gradient descent backpropagation	5	Training	0.83	0.69	0.60	0.04
		Validation	0.67	0.45	0.39	0.07
		Testing	0.65	0.42	0.57	0.11
		Overall	0.77	0.59	0.56	0.06
Resilient backpropagation	7	Training	**0.95**	**0.90**	**0.90**	**0.02**
		Validation	**0.95**	**0.90**	**0.91**	**0.04**
		Testing	**0.93**	**0.86**	**0.97**	**0.04**
		Overall	**0.95**	**0.90**	**0.90**	**0.03**
Supervised weight and bias algorithms						
Batch training with weight and bias learning rate	7	Training	0.80	0.64	0.59	0.10
		Validation	0.67	0.45	0.49	0.13
		Testing	0.76	0.58	0.57	0.11
		Overall	0.76	0.58	0.57	0.06
Random order weight and bias	10	Training	**0.89**	**0.79**	**0.82**	**0.06**
		Validation	**0.84**	**0.71**	**0.74**	**0.10**
		Testing	**0.88**	**0.77**	**1.10**	**0.06**
		Overall	**0.87**	**0.76**	**0.83**	**0.06**

Figure 3 shows the training, validation, testing and overall models of the three best algorithms developed using 7 neurons. Model 1 (Figure 3a), which was developed with the Levenberg-Marquardt algorithm, had a training $R = 0.96$, and validation, testing and overall $R = 0.95$, furthermore, the overall model had 6.86% of outliers according to the 95% confidence bounds. Figure 3b shows Model 2 with the Bayesian regularization algorithm with $R = 0.99$ for the training stage, $R = 0.98$ for testing and overall model with $R = 0.98$ and 5.88% outliers, this algorithm does not use a validation stage. On the other hand, Figure 3c depicts Model 3 developed using the RPROP algorithm, which also had a high $R = 0.95$ for training and validation stages, $R = 0.93$ for testing and an overall model with $R = 0.95$ and a low percentage of outliers (4.90%). It can be observed that in the overall models, some predicted values are >1 or < -1, this is because the targets were normalized based on the range of data obtained in the study (3–7); however, the liking hedonic scale is within the 1–9 range, therefore, a value <-1 or >1, will still fit within the 1–9 scale when reversing the normalization.

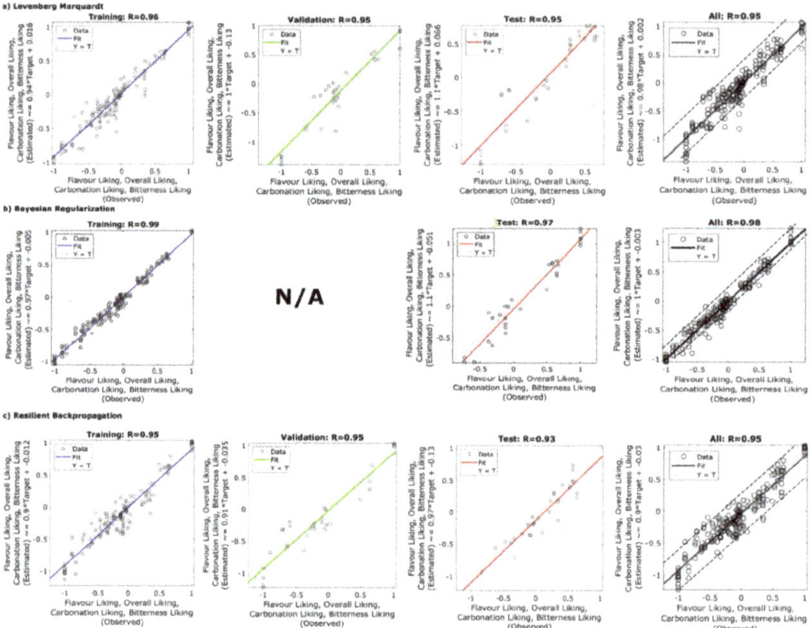

Figure 3. Models showing the three stages (training, validation and testing) as well as overall model of the three best algorithms found to assess liking of beer from morpho-colorimetric parameters from beer and beer foam: (**a**) Levenberg Marquardt, (**b**) Bayesian Regularization and (**c**) Resilient Backpropagation, showing the correlation coefficient (R) and 95% confidence bounds. In all graphs, the x-axis represents the observed data and y-axis the predicted or estimated values. N/A = not applicable.

4. Discussion

According to Beale, et al. [24], an indicator of a good model with no overfitting is when the validation correlation coefficient is close to the value from the training stage, which was met by the three best models found in this paper (Table 3 and Figure 3). The Bayesian regularization model (Model 2) does not have a validation stage; however, the R values of the other three stages are high and similar. Furthermore, an indication of a model with no overfitting is that the training performance (MSE) must be lower than the other stages, and the gap between the validation and testing MSE must be small [3,24]. This was also met by the best models found in this paper (Table 3).

The Levenberg-Marquardt algorithm (Model 1) is a backpropagation function, which works by calculating the second derivatives of a cost function. The advantages of this algorithm are: (i) that it is capable of giving a solution even though its start-point is far from the final minimum, (ii) its processing time is one of the lowest compared to other algorithms, (iii) the training algorithm stops when it finds the maximum epoch and (iv) the best performance value is achieved, or when it finds that the gradient value is lower than its minimum [25]. However, some disadvantages include: (i) it may not always secure a global optimum for an unrestrained optimization issue and ii) it may require higher memory usage [26]. On the other hand, the Bayesian regularization algorithm (Model 2) works using the same principles of Levenberg Marquardt but updating the weights and biases according to the optimization. The main advantages of this algorithm include: (i) lower memory usage, (ii) it has a good generalization for noisy or small datasets, (iii) it avoids overfitting effectively and (iv) it does not require a validation stage [17,25,27]. The RPROP (Model 3) works through an adaptation of the weight values according to the information of the local gradient, based only on the sign of the derivative. Its purpose is to avoid the negative effects of the small magnitude of partial derivatives which often result in small or null changes in weights and biases. The training stops when it reaches the maximum number of epochs or time, or when the best performance has been reached [28,29]. Some of the advantages of RPROP are: (i) the performance is better than other techniques used for adaptation [30] and (ii) it has fast convergence and low memory usage [31].

Based on the results from the three best models found to assess beer liking and acceptability by consumers, and considering the advantages and disadvantages of the algorithms, it can be said that Model 2 is the most appropriate for the prediction of beer liking using beer color and foam-related parameters. This is based on the highest correlation coefficient ($R = 0.98$), best performance, good fit within the confidence bounds with a low number of outliers, overall slope $b = 1$ and, therefore, no signs of overfitting. Furthermore, the dataset used met the small database requirements ($N = 51$), which is appropriate for the Bayesian Regularization.

The implementation of the models presented in this paper would allow a reduction in time and costs for the brewers when developing new products. It may also be used to do a fast-screening of any new developments without the need to conduct large sensory tests with consumers, which requires time for preparation, data gathering and analysis as well as financial resources for sampling and recruiting of consumers. This model allows accurate prediction of the liking of carbonation mouthfeel, flavor, bitterness, and overall liking using the physical parameters related to color and foam, this being possible because consumers are able to judge beer quality and acceptability based only on the visual attributes which give the first impression [8,9,32]. Furthermore, there is a relationship between the foam and color-related parameters, and bitterness as the iso-α-acids derived from hops are responsible for bitterness, but also contribute to foamability and foam stability due to their tensio-active properties. Furthermore, hops contribute to the development of aromas and flavors in beer, and foam aids in the release of aromas and flavors when bubbles burst [8,13,33,34].

Since the models are based on an automated data gathering process by using the RoboBEER and video analysis of pouring using computer vision algorithms, an artificial intelligence (AI) application may be implemented. This will offer to the beer industry a completely automated process to predict liking and acceptability of different beers by consumers.

5. Conclusions

The comparison of different artificial neural network algorithms aids in the selection of the best model making sure that it has no overfitting and it has the best performance. However, it is also important to consider the advantages and disadvantages of the algorithms in accordance with the dataset details and intended application to make the best choice. The best algorithm for the specific model presented in this paper was the Bayesian Regularization with very high accuracy ($R = 0.98$), and it would aid in the optimization of costs and time for breweries to assess beer acceptability without the need of recruiting consumers and running sensory sessions, being able to get the results within minutes. This is important, especially when having a large number of prototypes when developing new beer products. The use of the RoboBEER, computer vision algorithms and the ANN algorithms found in this research will allow the implementation of an AI system for the brewing industry to assess the effectiveness of beer making in terms of quality and acceptability of consumers.

Supplementary Materials: The following are available online at http://www.mdpi.com/2306-5710/5/2/33/s1, Table S1: Statistical results of the models developed using the backpropagation with Jacobian derivatives algorithm. Numbers in green and bold represent the models with the highest correlation and determination coefficients. Table S2: Statistical results of the models developed using the backpropagation with gradient derivative algorithms. Numbers in red and italics represent the models with the lowest correlation and determination coefficients, while those in green and bold represent the highest values. Table S3: Statistical results of the models developed using the supervised weight and bias algorithms. Numbers in red and italics represent the models with the lowest correlation and determination coefficients.

Author Contributions: Conceptualization, C.G.V., D.D.T. and S.F.; Formal analysis, C.G.V. and S.F.; Methodology, C.G.V. and S.F.; Supervision, D.D.T., F.R.D. and S.F.; Validation, C.G.V. and S.F.; Writing—original draft, C.G.V.; Writing—review & editing, D.D.T., F.R.D. and S.F.

Funding: This research received no external funding.

Acknowledgments: We gratefully acknowledge the support of NVIDIA Corporation with the donation of the Titan Xp GPU used for this research. This research was supported by the Australian Government through the Australian Research Council [Grant number IH120100053] "Unlocking the Food Value Chain: Australian industry transformation for ASEAN markets". C.G.V. is supported by the Melbourne Research Scholarship from the University of Melbourne.

Conflicts of Interest: The authors declare no conflict of interest.

References

1. Michalski, R.S.; Carbonell, J.G.; Mitchell, T.M. *Machine Learning: An Artificial Intelligence Approach*; Elsevier Science: Amsterdam, The Netherlands, 2014.
2. Bell, J. *Machine Learning: Hands-On for Developers and Technical Professionals*; Wiley: Hoboken, NJ, USA, 2014.
3. Goodfellow, I.; Bengio, Y.; Courville, A. *Deep Learning*; MIT Press: Cambridge, MA, USA, 2016.
4. Guyon, I.; Weston, J.; Barnhill, S.; Vapnik, V. Gene selection for cancer classification using support vector machines. *Mach. Learn.* **2002**, *46*, 389–422. [CrossRef]
5. Polat, K.; Güneş, S. Breast cancer diagnosis using least square support vector machine. *Digit. Signal Process.* **2007**, *17*, 694–701. [CrossRef]
6. Gonzalez Viejo, C.; Fuentes, S.; Li, G.; Collmann, R.; Condé, B.; Torrico, D. Development of a robotic pourer constructed with ubiquitous materials, open hardware and sensors to assess beer foam quality using computer vision and pattern recognition algorithms: RoboBEER. *Food Res. Int.* **2016**, *89*, 504–513. [CrossRef]
7. Gonzalez Viejo, C.; Fuentes, S.; Torrico, D.; Howell, K.; Dunshea, F.R. Assessment of beer quality based on foamability and chemical composition using computer vision algorithms, near infrared spectroscopy and machine learning algorithms. *J. Sci. Food Agric.* **2018**, *98*, 618–627. [CrossRef] [PubMed]
8. Gonzalez Viejo, C.; Fuentes, S.; Howell, K.; Torrico, D.D.; Dunshea, F.R. Integration of non-invasive biometrics with sensory analysis techniques to assess acceptability of beer by consumers. *Physiol. Behav.* **2019**, *200*, 139–147. [CrossRef]
9. Gonzalez Viejo, C.; Fuentes, S.; Howell, K.; Torrico, D.; Dunshea, F.R. Robotics and computer vision techniques combined with non-invasive consumer biometrics to assess quality traits from beer foamability using machine learning: A potential for artificial intelligence applications. *Food Control* **2018**, *92*, 72–79. [CrossRef]
10. Fuentes, S.; Hernández-Montes, E.; Escalona, J.; Bota, J.; Gonzalez Viejo, C.; Poblete-Echeverría, C.; Tongson, E.; Medrano, H. Automated grapevine cultivar classification based on machine learning using leaf morpho-colorimetry, fractal dimension and near-infrared spectroscopy parameters. *Comput. Electron. Agric.* **2018**, *151*, 311–318. [CrossRef]
11. Romero, M.; Luo, Y.; Su, B.; Fuentes, S. Vineyard water status estimation using multispectral imagery from an UAV platform and machine learning algorithms for irrigation scheduling management. *Comput. Electron. Agric.* **2018**, *147*, 109–117. [CrossRef]
12. Yu, H.Y.; Niu, X.Y.; Lin, H.J.; Ying, Y.B.; Li, B.B.; Pan, X.X. A feasibility study on on-line determination of rice wine composition by Vis–NIR spectroscopy and least-squares support vector machines. *Food Chem.* **2009**, *113*, 291–296. [CrossRef]
13. Gonzalez Viejo, C.; Fuentes, S.; Torrico, D.D.; Howell, K.; Dunshea, F.R. Assessment of Beer Quality Based on a Robotic Pourer, Computer Vision, and Machine Learning Algorithms Using Commercial Beers. *J. Food Sci.* **2018**, *83*, 1381–1388. [CrossRef] [PubMed]

14. Ellis, D.I.; Broadhurst, D.; Kell, D.B.; Rowland, J.J.; Goodacre, R. Rapid and quantitative detection of the microbial spoilage of meat by Fourier transform infrared spectroscopy and machine learning. *Appl. Environ. Microbiol.* **2002**, *68*, 2822–2828. [CrossRef]

15. Mathworks Inc. *Mastering Machine Learning: A Step-by-Step Guide with MATLAB*; Mathworks Inc.: Sherborn, MA, USA, 2018.

16. Lin, M.-I.B.; Groves, W.A.; Freivalds, A.; Lee, E.G.; Harper, M. Comparison of artificial neural network (ANN) and partial least squares (PLS) regression models for predicting respiratory ventilation: An exploratory study. *Eur. J. Appl. Physiol.* **2012**, *112*, 1603–1611. [CrossRef] [PubMed]

17. Amini, M.; Abbaspour, K.C.; Khademi, H.; Fathianpour, N.; Afyuni, M.; Schulin, R. Neural network models to predict cation exchange capacity in arid regions of Iran. *Eur. J. Soil Sci.* **2005**, *56*, 551–559. [CrossRef]

18. Schaap, M.G.; Leij, F.J.; Van Genuchten, M.T. Neural network analysis for hierarchical prediction of soil hydraulic properties. *Soil Sci. Soc. Am. J.* **1998**, *62*, 847–855. [CrossRef]

19. Ogunoiki, A.; Olatunbosun, O. *Artificial Road Load Generation Using Artificial Neural Networks*; 0148-7191; SAE Technical Paper; SAE: Warrendale, PA, USA, 2015.

20. Buss, D. Food Companies Get Smart About Artificial Intelligence. *Food Technol.* **2018**, *72*, 26–41.

21. Cajka, T.; Riddellova, K.; Tomaniova, M.; Hajslova, J. Recognition of beer brand based on multivariate analysis of volatile fingerprint. *J. Chromatogr. A* **2010**, *1217*, 4195–4203. [CrossRef]

22. Iñón, F.A.; Garrigues, S.; de la Guardia, M. Combination of mid-and near-infrared spectroscopy for the determination of the quality properties of beers. *Anal. Chim. Acta* **2006**, *571*, 167–174. [CrossRef]

23. Gonzalez Viejo, C.; Fuentes, S.; Torrico, D.; Lee, M.; Hu, Y.; Chakraborty, S.; Dunshea, F. The Effect of Soundwaves on Foamability Properties and Sensory of Beers with a Machine Learning Modeling Approach. *Beverages* **2018**, *4*, 53. [CrossRef]

24. Beale, M.H.; Hagan, M.T.; Demuth, H.B. *Deep Learning Toolbox User's Guide*; Mathworks Inc.: Sherborn, MA, USA, 2018.

25. Markopoulos, A.P.; Georgiopoulos, S.; Manolakos, D.E. On the use of back propagation and radial basis function neural networks in surface roughness prediction. *J. Ind. Eng. Int.* **2016**, *12*, 389–400. [CrossRef]

26. Saduf, M.A.W. Comparative study of back propagation learning algorithms for neural networks. *Int. J. Adv. Res. Comput. Sci. Softw. Eng.* **2013**, *3*, 1151–1156.

27. Kayri, M. Predictive abilities of bayesian regularization and Levenberg–Marquardt algorithms in artificial neural networks: A comparative empirical study on social data. *Math. Comput. Appl.* **2016**, *21*, 20. [CrossRef]

28. Riedmiller, M.; Braun, H. A Direct Adaptive Method for Faster Backpropagation Learning: The RPROP Algorithm. In Proceedings of the IEEE International Conference on Neural Networks, San Francisco, CA, USA, 28 March–1 April 1993; pp. 586–591.

29. Mathworks Inc. Resilient Backpropagation. Available online: https://au.mathworks.com/help/deeplearning/ref/trainrp.html (accessed on 1 October 2018).

30. Patnaik, L.M.; Rajan, K. Target detection through image processing and resilient propagation algorithms. *Neurocomputing* **2000**, *35*, 123–135. [CrossRef]

31. Pajchrowski, T.; Zawirski, K.; Nowopolski, K. Neural speed controller trained online by means of modified RPROP algorithm. *IEEE Trans. Ind. Inform.* **2015**, *11*, 560–568. [CrossRef]

32. Bamforth, C. Perceptions of beer foam. *J. Inst. Brew.* **2000**, *106*, 229–238. [CrossRef]

33. De Keukeleire, D. Fundamentals of beer and hop chemistry. *Quim. Nova* **2000**, *23*, 108–112. [CrossRef]

34. Liger-Belair, G.; Cilindre, C.; Gougeon, R.D.; Lucio, M.; Gebefügi, I.; Jeandet, P.; Schmitt-Kopplin, P. Unraveling different chemical fingerprints between a champagne wine and its aerosols. *Proc. Natl. Acad. Sci. USA* **2009**, *106*, 16545–16549. [CrossRef]

Article

The Effect of Soundwaves on Foamability Properties and Sensory of Beers with a Machine Learning Modeling Approach

Claudia Gonzalez Viejo, Sigfredo Fuentes *, Damir D. Torrico, Mei Huii Lee, Yue Qin Hu, Sanjit Chakraborty and Frank R. Dunshea

School of Agriculture and Food, Faculty of Veterinary and Agricultural Sciences, University of Melbourne, Melbourne, VIC 3010, Australia; cgonzalez2@unimelb.edu.au (C.G.V.); damir.torrico@unimelb.edu.au (D.D.T.); meil1@student.unimelb.edu.au (M.H.L.); yueh2@student.unimelb.edu.au (Y.Q.H.); sanjitc@student.unimelb.edu.au (S.C.); fdunshea@unimelb.edu.au (F.R.D.)
* Correspondence: sfuentes@unimelb.edu.au; Tel.: +61-3-9035-9670

Received: 10 July 2018; Accepted: 25 July 2018; Published: 26 July 2018

Abstract: The use of ultrasounds has been implemented to increase yeast viability, de-foaming, and cavitation in foods and beverages. However, the application of low frequency audible sound to decrease bubble size and improve foamability has not been explored. In this study, three treatments using India Pale Ale beers were tested, which include (1) a control, (2) the application of audible sound during fermentation, and (3) the application of audible sound during natural carbonation. Five different audible frequencies (20 Hz, 30 Hz, 45 Hz, 55 Hz, and 75 Hz) were applied daily for one minute each (starting from the lowest frequency) during fermentation (11 days, treatment 2) and carbonation (22 days, treatment 3). Samples were measured in triplicates using the RoboBEER to assess color and foam-related parameters. A trained panel ($n = 10$) evaluated the intensity of sensory descriptors. Results showed that samples with sonication treatment had significant differences in the number of small bubbles, alcohol, and viscosity compared to the control. Furthermore, except for foam texture, foam height, and viscosity, there were non-significant differences in the intensity of any sensory descriptor, according to the rating from the trained sensory panel. The use of soundwaves is a potential treatment for brewing to improve beer quality by increasing the number of small bubbles and foamability without disrupting yeast or modifying the aroma and flavor profile.

Keywords: foamability; audible sound; brewing; carbonation; fermentation

---

1. Introduction

Sound consists of a mechanical longitudinal wave, which propagates through gas and liquid media such as air and water but may also do so through solid materials [1]. Furthermore, frequency is defined as the number of complete oscillations per second and is expressed in Hertz [2,3]. The wavelength is the distance between the peaks of the sound signal [4] and is calculated using the following equation: $V = \lambda F$, where "V" is the speed of sound, λ is the wavelength, and "F" is the frequency [5]. The wavelength depends on the velocity. Therefore, the medium affects the wavelength since the density and elasticity of the medium will change the velocity of sound [6]. Soundwaves are divided in three different classes according to their frequency range including (i) infrasound, which are those below 20 Hz and not audible by the human ear but may be detected by measuring the variation in pressure and vibrations. This type of soundwaves is naturally produced by earthquakes, volcanic eruptions, among others. A second class of soundwaves includes (ii) audible sound, which is produced within 20 Hz and 2000 Hz and, as its name implies, it can be perceived by the human ear and (iii) ultrasound that is above 2000 Hz and cannot be heard by humans. However, some animals are able to hear it [7–9]. In food-related products, infrasound has been used to reduce membrane fouling in beer and wine filtration [10] while audible sound has been used to increase *Escherichia coli* growth [11], in the control of insects in stored products such as grains [12], and to increase yeast cells growth rate [13] among others. However, ultrasound has been the most applied frequency range in food and beverages to increase the retention of nutritional compounds in fruit beverages [14], to increase production capacity and promote lightness of juices, and to aid in their preservation by the inactivation of enzymes [15] for the degassing of liquids [16] among others.

The amount and stability of foam and bubbles are essential in carbonated beverages such as beer since they comprise the most important factors that consumers consider when assessing the quality of beer. According to previous studies, consumers have a preference for beers with a medium level of foam height and consider the low foam as non-desirable with the lowest liking, lowest perceived quality scores, and highest penalty scores [17–21]. Therefore, it is important to explore methods such as sonication during its production process to increase foam in beer. However, the application of soundwave frequencies in beer has been limited to infra-sound or ultrasound with other purposes such as the aforementioned reduction of membrane fouling during filtration [10], to measure density of beer during fermentation [22], hops extraction [23], to increase yeast viability, and to increase alcohol concentration applying ultrasound during fermentation [24] among others.

The use of soundwaves to modify bubble size has been narrowed to the purpose of degassing, de-foaming [16], and to produce cavitation, which is the effect of forming bubbles that increase their size and cause their implosion [25]. Nevertheless, there are no studies that had assessed the effects of audible sound in bubbles and foam quality in carbonated beverages. Some breweries such as Philadelphia's Dock Street Brewery, Mikkeller craft brewery, and Garage Project have applied music during the fermentation stage of the brewing process by claiming that music improves beers sensory characteristics and increases yeast activity [26–28]. However, music has a mix of different frequencies and soundwaves, which does not allow us to isolate the effects of the specific frequency levels. This can potentially have a higher, lower, or no effect on beer characteristics such as foaming and bubble size.

This paper aims to present the results from the assessment of the effects of audible soundwaves on beer bubbles and foam quality during fermentation and natural carbonation stages of the brewing process. The study was conducted using triplicates of three different treatments including (i) a control using the usual brewing process, (ii) the application of frequencies during fermentation, and (iii) the application of frequencies during the carbonation stage. Five different frequency levels (20 Hz, 30 Hz, 45 Hz, 55 Hz, and 75 Hz) were used for the treatments. These were applied using two speakers including an amplifier and an iPhone application Audio Function Generator (Thomas Gruber, Forchenstein, Austria). The samples were analyzed using a robotic pourer RoboBEER and Matlab® R2018b (Mathworks, Inc., Matick, MA, USA) to assess foam and color-related parameters and through a trained sensory panel ($n = 10$) to assess significant differences between the treatments in their sensory descriptors. Lastly, two machine learning models previously developed by Gonzalez Viejo et al. [29,30] were tested by feeding the inputs from the RoboBEER of the triplicates of the three treatments to (i) predict the type of fermentation and (ii) to predict the intensities of sensory descriptors by obtaining high accuracy in the testing.

2. Materials and Methods

2.1. Beer Samples Description and Processing

English style India Pale Ale (IPA, Berlin IPA, BrewBaker, Berlin, Germany) samples were selected for this study since breweries who have used music for brewing have done it using this style of beer [26,28]. The samples were brewed using the PicoBrew S (PicoBrew, Seattle, WA, USA). The samples came in special containers specific for the PicoBrew, which included the malted barley and different types of hops (Perle, Polaris, Tettnanger, Smaragd, and Cascade). As shown in Figure 1, for the control and the other two treatments (soundwaves during fermentation = SWF and soundwaves during carbonation = SWC), the container including the ingredients was inserted in the PicoBrew machine and followed the default formula in the machine for the specified IPA sample (Bitterness = 65 IBU). The first part of the brewing process for the control and two treatments lasted 2.5 h and consisted of heating, doughing, mash 1, mash 2, mash out, boiling to a maximum temperature of 114 °C, and holding at 94 °C for 1 h with the addition of four different types of hops at the start, the middle, and the end of the boiling process (Figure 1 (1)). Following the first part of the process, the yeast (*Saccharomyces cerevisiae*; strain 1056; Safale US-05, Belgium) was added to the three treatments in the same way. Then the hermetically sealed kegs for the control and SWC were left at room temperature (25 °C) in the same place for 11 days while the SWF, which was also hermetically sealed, was treated with audible soundwaves daily by applying frequencies of 20 Hz, 30 Hz, 45 Hz, 55 Hz, and 75 Hz at −4 dB for 1 min each during 11 days of fermentation. The soundwaves were applied using two speakers, an amplifier, an iPhone 5s (Apple Inc., Cupertino, CA, USA), and an Audio Function Generator application (Thomas Gruber) (Figure 1 (2)). To make sure the fermentation was completed, the CO_2 from the keg was released once and then tried again 1 h later to make sure there was no more gas production. The last part of the process consisted of bottling the samples in brown bottles and adding one sugar drop for the natural carbonation process in which all the samples were left at 25 °C for 22 days. The control and SWF were left in the sample place during this stage of the process while the SWC were treated with audible soundwaves every day by applying frequencies of 20 Hz, 30 Hz, 45 Hz, 55 Hz, and 75 Hz at −4 dB for 1 min each for 22 days (Figure 1 (3)).

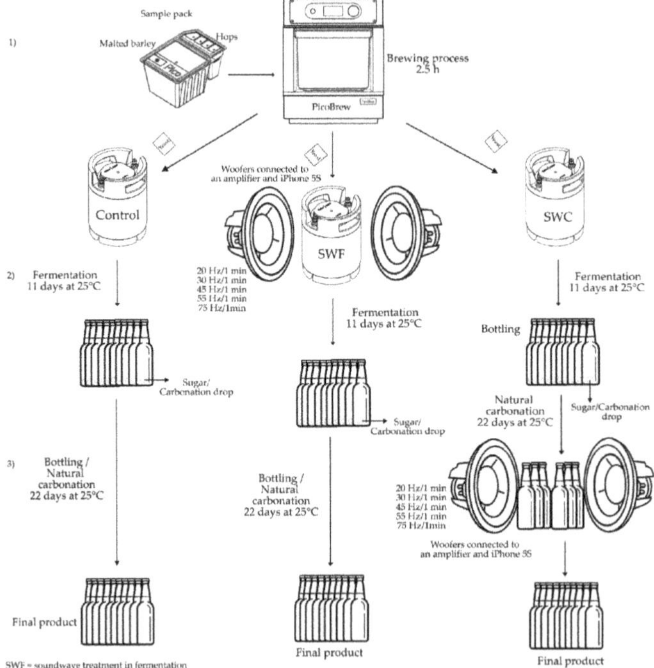

Figure 1. Diagram representing the brewing process and soundwave treatments applied to the different beer samples such as Control, SWF = soundwaves application at fermentation, and SWC = soundwaves applied in the carbonation process. Where (1) brewing process, (2) fermentation, and (3) bottling and natural carbonation.

The specific frequencies were selected by recording videos of the application of different levels (20–100 Hz) at −4 dB of amplitude to milk and dry thyme leaves in a Petri dish (Figure 2) and were further magnified using the Eulerian Magnification Algorithm [31] in Matlab® R2018b (Mathworks, Inc., Matick, MA, USA). Those frequencies that produced movements of the thyme in different directions such as grouping particles, separating them, and vibrating in the same position were selected (20 Hz, 30 Hz, 45 Hz, 55 Hz, and 75 Hz). Dry thyme and milk were used due to the lack of solubility and to ease the magnification and visibility of particle movement.

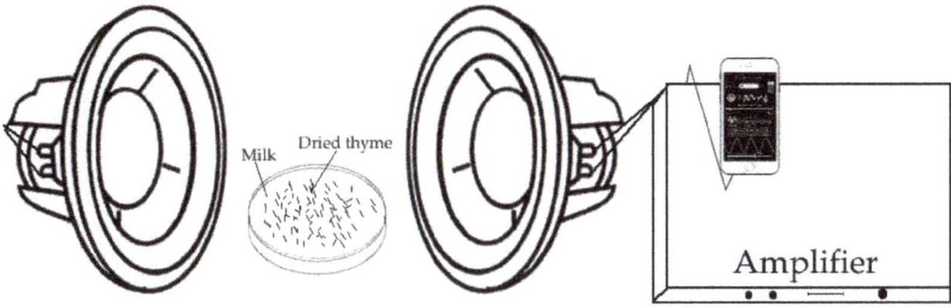

Figure 2. Diagram representing the setup of the testing of different audible sound frequencies to select the most appropriate ones for the treatments.

2.2. Color and Foam-Related Parameters

All samples were analyzed in triplicates using computer vision algorithms written in Matlab® R2018b (Mathworks, Inc., Matick, MA, USA) to analyze 5-min videos recorded while using the automatic robotic pourer RoboBEER to obtain 15 color and foam-related parameters including: (i) maximum volume of foam (MaxVol), (ii) total lifetime of foam (TLTF), (iii) lifetime of foam (LTF), (iv) foam drainage (FDrain), color in two scales (v), (vi), (vii) CIELab and (viii), (ix), (x) RGB and bubble size distribution of (xi) small (SmBubb), (xii) medium (MedBubb), and (xiii) large bubbles (LgBubb) as well as (xiv) alcohol gas (OH) and (xv) carbon dioxide (CO_2) release. The samples were poured in the same International Standard Wine Tasting Glass Luigi Bormioli to avoid differences due to the glass. Details about the procedure of the RoboBEER can be found in the paper published by Gonzalez Viejo et al. [29]. The three parameters from the RGB color scale were converted to a color index using Equations (1)–(4). The samples were also analyzed for alcohol content in the liquid (Alcohol) using an Alcolyzer Wine M alcohol-meter (Anton Paar GmbH, Graz, Austria). Additionally, pH was measured using a pH-meter Benchtop pH/mV meter 860031 (Sper scientific direct, Scottsdale, AZ, USA) with 50 mL of the sample at an ambient temperature (25 °C). Furthermore, viscosity (Visc) was measured with a Brookfield viscometer DV-II+ (AMETEK Brookfield, Middleborough, MA, USA) using 150 mL of the sample and an RV02 spindle at 50 rpm for 20 s.

$$RGB\ Intensity\ ratio(I) = R + G + B \qquad (1)$$

$$Red\ intensity\ ratio\ (RI) = \frac{R}{I} \qquad (2)$$

$$Green\ intensity\ ratio\ (GI) = \frac{G}{I} \qquad (3)$$

$$Blue\ intensity\ ratio\ (BI) = \frac{B}{I} \qquad (4)$$

2.3. Sensory Descriptive Analysis

A sensory session was conducted using a trained panel of 10 participants who were regular beer consumers using the quantitative descriptive analysis (QDA) method. Two 1.5 h training sessions using IPA beers were conducted previous to the evaluation of the samples to obtain the descriptors and for the participants to familiarize themselves with the test. The sensory session was conducted at room temperature (24 °C) in a focus room in the sensory laboratory of the Faculty of Veterinary and Agricultural Sciences of The University of Melbourne, Australia (FVAS-UoM). Triplicates of the samples were evaluated and all were served in 1 oz plastic glasses at a refrigerated temperature (4 °C). The tasting was double blind and all samples were labeled with 3-digit random codes. Table 1 shows the parameters evaluated for all samples and the anchors used for each. For the visual assessment, the first 20 s of the videos obtained using the RoboBEER were shown on a large screen to all participants at the same time to ensure uniformity in the pouring and to evaluate all samples under the same conditions. The questionnaire was displayed in a Samsung Galaxy View Tablet (Samsung Group, Seoul, South Korea) using the Bio-sensory application [17,18,32,33]. All descriptors were rated using a 15 cm non-structured scale.

Table 1. Descriptors, abbreviation, and anchors used for the sensory descriptive test.

Descriptor	Abbreviation	Anchors
Foam Stability	FStab	Short time–Long time
Foam Height	FHeight	Short–High
Foam Texture (Bubble size)	FText	Small–Large
Color Intensity	CInt	Light–Dark
Clarity	Clarity	Haze–Clear
Aroma–Hops	AHops	Absent–Intense
Aroma–Spices	ASpices	Absent–Intense
Aroma–Floral	AFloral	Absent–Intense
Aroma–Fruity	AFruity	Absent–Intense
Aroma–Brown Sugar	ABSugar	Absent–Intense
Aroma–Yeast	AYeast	Absent–Intense
Aroma–Nuts	ANut	Absent–Intense
Aroma–Grains	AGrain	Absent–Intense
Viscosity	MVisc	Thin–Thick
Astringency	MAstr	Absent–Intense
Carbonation Mouthfeel	MCarb	Absent–Intense
Warming Mouthfeel	MWarm	Absent–Intense
Taste–Bitter	TBitt	Absent–Intense
Taste–Sweet	TSweet	Absent–Intense
Taste–Sour	TSour	Absent–Intense
Flavor–Hops	FHops	Absent–Intense

2.4. Statistical Analysis

Data obtained from the RoboBEER were analyzed using multivariate data analysis based on the PCA with a customized code written in Matlab® R2018b. The factor loadings are shown as supplementary material (Table S1). A correlation matrix (CM) was developed in Matlab® R2018b to assess significant correlations ($p < 0.05$). Furthermore, all data were assessed for significant differences using ANOVA and the least significant differences (LSD) post-hoc test ($\alpha = 0.05$) in SAS® 9.4 software (SAS Institute Inc., Cary, NC, USA). Mean and standard deviation (SD) values were obtained.

Two machine learning models developed using commercial beers with artificial neural networks (ANN) [29,30] were fed with the 15 parameters obtained from the RoboBEER for the control. SWF and SWC were used as inputs to predict the type of fermentation and to predict the intensity of ten sensory descriptors (AHops, AYeast, AGrain, MVisc, MAstr, MCarb, TBitt, TSweet, TSour, and FHops). Results were correlated with those obtained using the trained sensory panel to validate the accuracy of the model for sensory descriptors. To assess the later, a Pearson linear correlation ($y = ax + b$) was developed, Statistical data such as the correlation coefficient (R), determination coefficient (R^2), and slope were obtained.

3. Results

3.1. Color and Foam-Related Parameters

Figure 3a shows the principal components analysis (PCA) of the foam and color-related parameters and CO_2 and alcohol gas release were obtained using the RoboBEER. It can be observed that the principal component one (PC1) represented 50.73% of data variability while principal component two (PC2) accounted for 21.05% with the PCA explaining a total of 71.78%. According to the factor loadings (Table S1), the PC1 was mainly represented by the TLTF, MaxVol, LTF, and color parameter a in the positive side and by GI and L on the negative side of the axis. On the other hand, the PC2 was primarily represented by BI and L on the positive side and by RI and a color parameter b on the negative side of the axis. It is shown that the control samples were more represented by the FDrain and green and yellow colors (GI and b) while two of the replicates of SWC had more GI, L, and BI. The third replicate was more characterized by small, medium, and large bubbles and a red color (a and RI). The SWF samples were more represented by the foam-related parameters such as MaxVol, LTF, TLTF, SmBubb, LgBubb, MedBubb, and CO_2.

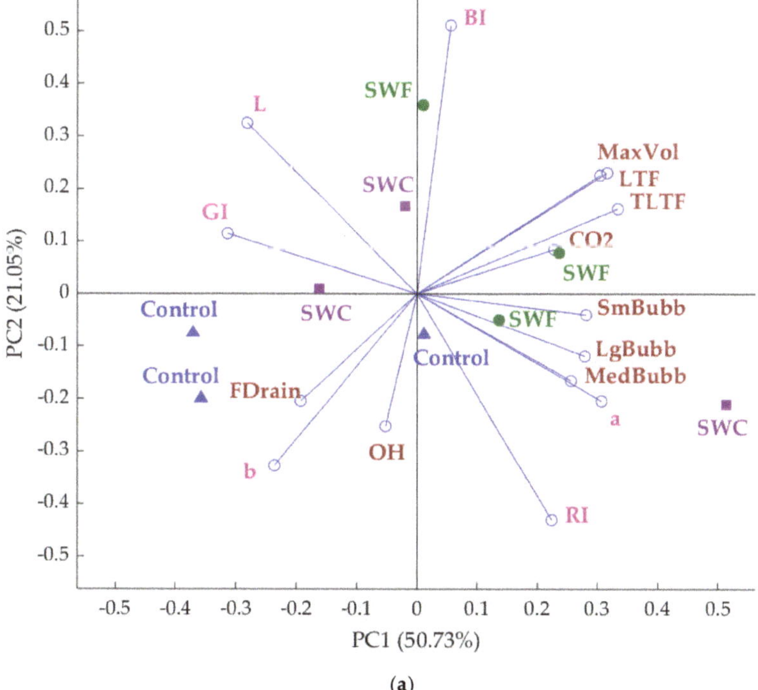

Figure 3. *Cont.*

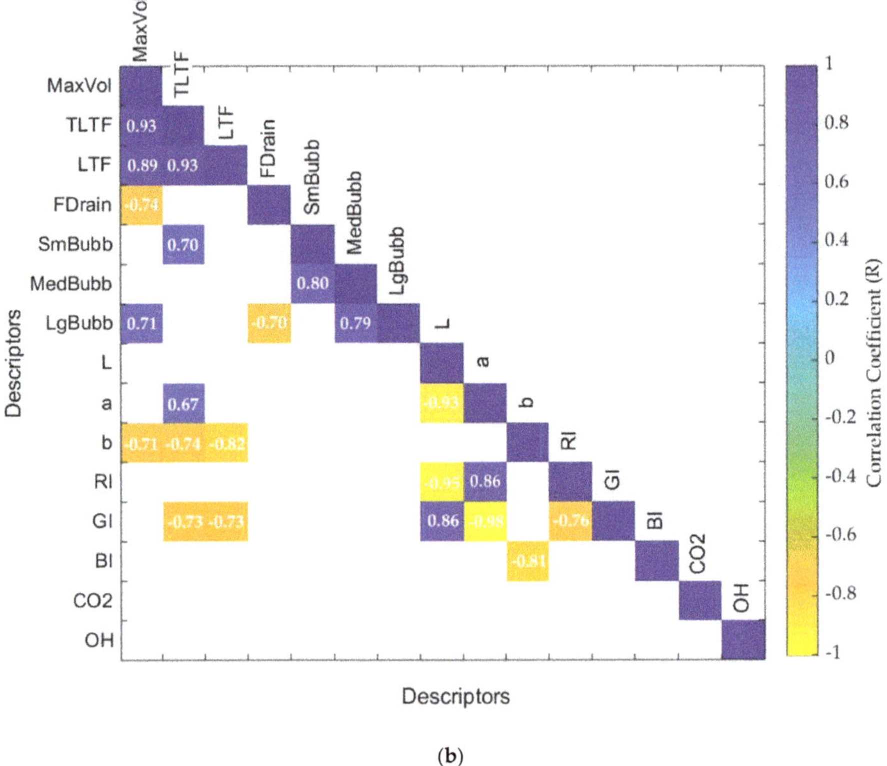

(b)

Figure 3. Results from multivariate data analysis showing: (**a**) principal component analysis where the x-axis represents the principal component one (PC1) and y-axis represents the principal component two (PC2), samples in blue and triangles represent the control, those in green, and circles the treatment of soundwaves during fermentation (SWF) and the purple squares the soundwaves during carbonation (SWC), and (**b**) the correlation matrix shows only the significant correlations ($p < 0.05$) where the color bar depicts the positive (blue) and negative (yellow) correlations between the different descriptors. The values inside the boxes are the correlation coefficients (R). Abbreviations: MaxVol = maximum volume of foam, TLTF = total lifetime of foam, LTF = lifetime of foam, FDrain = foam drainage, L, a, b = color in CIELab scale, RI, GI, and BI indices of color in the RGB scale. SmBubb = small bubbles. MedBubb = medium bubbles. LgBubb = large bubbles. OH = alcohol gas and CO_2 = carbon dioxide release.

Figure 3b shows the CM in which MaxVol had positive and significant correlation with TLTF ($R = 0.93$), TLF ($R = 0.89$), LgBubb (0.71), and a negative correlation with FDrain ($R = -0.74$). Furthermore, TLTF had a positive correlation with SmBubb ($R = 0.70$). Yet, as expected, the red color from the RGB scale (RI) had a positive correlation with a red color (a) from the Lab scale ($R = 0.86$). Likewise, there was a negative correlation between the blue color (BI) from the RGB scale and b from the Lab scale ($R = -0.81$).

Table 2 shows the results of the ANOVA for the foam-related parameters measured with the RoboBEER. It can be observed that there were significant differences in the number of small bubbles in the foam with SWC being the highest and significantly different from the control. While there were no significant differences for the other foam-related parameters, the MaxVol had a p-value of 0.055, which is close to the significant value. According to the SD, the SWF treatment was the most consistent in MaxVol, TLTF, LTF, FDrain, and LgBubb since it had the lowest values compared to the control and SWC.

Table 2. Means and standard deviation (SD) of the three treatments for the foam-related parameters.

Treatment	MaxVol (mL)	TLTF(s)	LTF (mL s^{-1})	FDrain (mL s^{-1})	SmBubb (number)	MedBubb (number)	LgBubb (number)
Control	25.7 [a] ± 14.0	1647.6 [a] ± 1288.5	594.9 [a] ± 537.7	42.4 [a] ± 16.4	597.2 [b] ± 390.0	8.3 [a] ± 9.9	4.0 [a] ± 4.3
SWC	43.9 [a] ± 12.9	3243.7 [a] ± 1468.0	1552.3 [a] ± 1222.3	36.2 [a] ± 9.2	2519.5 [a] ± 1651.5	28.2 [a] ± 49.9	5.7 [a] ± 4.9
SWF	39.2 [a] ± 9.4	3005.5 [a] ± 1139.4	1100.7 [a] ± 482.8	34.4 [a] ± 3.1	1400.0 [ab] ± 1030.2	7.8 [a] ± 11.3	4.5 [a] ± 3.3

Abbreviations: MaxVol = maximum volume of foam, TLTF = total lifetime of foam, LTF = lifetime of foam, FDrain = foam drainage, SmBubb = small bubbles, MedBubb = medium bubbles, LgBubb = large bubbles, SWC = application of soundwaves during carbonation, and SWF = application of soundwaves during fermentation. [a–b]: Different letters depict significant differences using the least significant difference (LSD) post-hoc test.

Table 3 shows the means and results from the ANOVA and LSD tests for the color parameters, alcohol, and CO_2. There were no significant differences in most of the parameters except for the +b (yellow color) being the highest value and the blue color index (BI) with SWC being the highest in this parameter. There were also significant differences between the SWF, which had the highest value as well as significant differences between the control and SWC for the alcohol content in the liquid and viscosity (Visc).

Table 3. Means and standard deviation (SD) of the three treatments for the color and analytical measurements.

Treatment	* L	* a	* b	* RI	* GI	* BI	* OH	CO_2 (ppm)	Alcohol (%)	* pH	Visc (cP)
Control	74.7 [a] ± 1.6	−3.9 [a] ± 1.4	68.6 [a] ± 1.2	0.6 [a] ± 0.0	0.4 [a] ± 0.0	0.02 [b] ± 0.0	287.0 [a] ± 16.6	12,055.6 [a] ± 10514.7	5.3 [b] ± 0.2	4.25 [a] ± 0.03	11.2 [b] ± 1.0
SWC	73.8 [a] ± 2.9	−2.0 [a] ± 1.9	63.2 [b] ± 2.5	0.6 [a] ± 0.0	0.4 [a] ± 0.0	0.04 [a] ± 0.01	286.8 [a] ± 43.6	16,020.5 [a] ± 13469.6	5.2 [b] ± 0.3	4.26 [a] ± 0.03	11.5 [b] ± 0.7
SWF	73.6 [a] ± 3.8	−2.5 [a] ± 2.8	65.5 [b] ± 2.8	0.6 [a] ± 0.0	0.4 [a] ± 0.0	0.03 [ab] ± 0.01	288.3 [a] ± 37.1	16,853.2 [a] ± 11429.4	5.6 [a] ± 0.1	4.28 [a] ± 0.02	12.7 [a] ± 0.6

* Unitless parameters. Abbreviations: L, a, b = color in CIELab scale, RI, GI, BI indices of color in RGB scale, OH = alcohol gas and CO_2 = carbon dioxide release, Visc = viscosity, SWC = application of soundwaves during carbonation, and SWF = application of soundwaves during fermentation. [a–b]: Different letters depict significant differences using the least significant difference (LSD) post-hoc test.

3.2. Descriptive Sensory Evaluation

Results from the descriptive sensory evaluation are found in Figure 4. It can be observed that there were non-significant differences in most of the sensory descriptors except for FHeight, FText, and MVisc. The sample rated as the highest in foam (FHeight) and foam texture (FText) was SWC, which is significantly different from and followed by SWF and control. Yet, SWF was the highest in viscosity (MVisc), which was similar to the control and with significant differences with SWC. Although there were no significant differences in AFruity, AFloral, and AGrain, the control and SWF had a higher rating than SWC. The three samples were rated high in bitter taste with ratings between 8.6 and 8.9 in a 15 cm intensity scale (Figure 4).

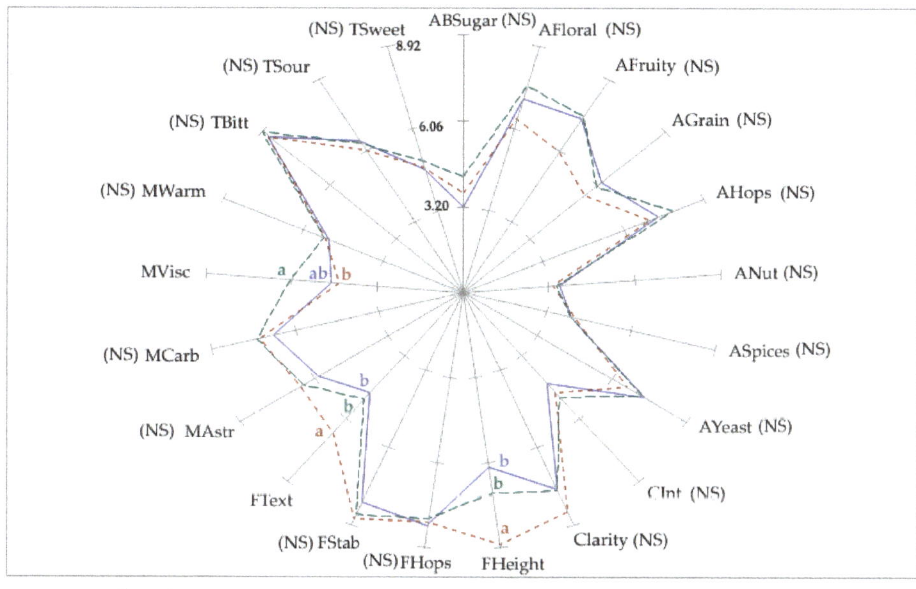

Figure 4. Spider chart showing the significant differences found for the intensity of sensory descriptors using the least significant difference (LSD) post-hoc test where NS represents non-significant differences and different letters depict significant differences. The numbers in the chart represent the mean values at that position of the scale. SWC = application of soundwaves during carbonation, and SWF = application of soundwaves during fermentation. Abbreviations of the descriptors can be found in Table 1.

3.3. Validation of Machine Learning Models

The results from the 15 foam and color-related parameters obtained using the RoboBEER were fed as inputs in an ANN model developed using commercial beers, which had a 92% accuracy as shown by Gonzalez Viejo et al. [29] to predict the type of fermentation. The results obtained showed that the three samples were classified as top fermentation (Table 4), which is accurate since IPA beers are brewed using top fermenting yeast at an ambient temperature (~25 °C). However, the three treatments especially SWC presented a few characteristics from spontaneous fermentation beers.

Table 4. Results from the classification of the type of fermentation using an artificial neural network model was developed using the color and foam-related parameters.

Treatment	Top	Bottom	Spontaneous
Control	0.996	0.000	0.004
SWF	0.997	0.000	0.003
SWC	0.994	0.000	0.006

Abbreviations: SWC = application of soundwaves during carbonation and SWF = application of soundwaves during fermentation.

Figure 5 shows the results from the validation of the second ANN model, which was also fed using the 15 foam and color-related parameters obtained using the RoboBEER to predict 10 of the sensory descriptors (AHops, AYeast, AGrain, MVisc, MAstr, MCarb, TBitt, TSweet, TSour, and FHops). The model used to predict the values had a correlation of $R = 0.91$ and a determination coefficient $R^2 = 0.83$. It can be found from Gonzalez Viejo et al. [30]. For the samples developed in this study, the correlation between the observed values obtained with the trained sensory panel and the predicted values using the model was $R = 0.85$ with a determination coefficient $R^2 = 0.72$ and a slope of 1.09.

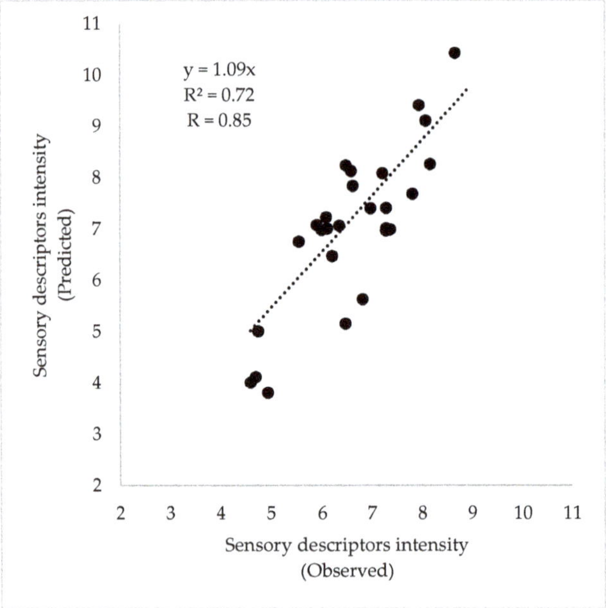

Figure 5. Correlation of the observed values (x-axis) of 10 sensory descriptors (AHops, AYeast, AGrain, MVisc, MAstr, MCarb, TBitt, TSweet, TSour, and FHops) obtained using the trained sensory panel and the predicted values (y-axis) using an artificial neural network model developed using the color and foam-related parameters.

4. Discussion

4.1. Color and Foam-Related Parameters

The negative correlation found between MaxVol and FDrain (Figure 3b) was expected since the FDrain is the excess of liquid drained by gravity from the wet foam to produce dry foam [34]. Therefore, the volume of foam decreases and the liquid volume increases [29]. In the PCA (Figure 3a), it was observed that there was a separation of the triplicates since one bottle of each had different characteristics. However, for the SWF, the separation of the triplicates in the PCA and the SD of the means (Tables 2 and 3) for most of the parameters was lower than the other treatments. Therefore, this might be due to the soundwaves treatment applied during the fermentation process, which could make the sample more uniform. According to Choi et al. [24], there is an increase in yeast viability when applying ultrasound during the fermentation process and, therefore, an increase in alcohol content in the beer. A similar effect was found in the present study, but, using audible sound, higher and significantly different values were found between the SWF and the other treatments in alcohol content in the liquid. Although non-significant, the MaxVol, TLTF, LTF, and SmBubb tended to be higher for the SWC and SWF when compared to the control. Significant differences were found for viscosity with SWF being the different treatment with the highest value followed by SWC. The viscosity of beer, which is mainly defined by the amount of surfactant substances such as proteins and carbohydrates, is directly linked with foam stability [35,36]. According to previous research, an ultrasound has been used for de-foaming since it causes the foam to collapse as the ultra-high frequencies provoke the coalescence of the bubbles, which causes bubbles breakage [37]. Another effect of ultrasound is to increase bubble size until it implodes (cavitation) [25]. Therefore, according to the findings in the present study, the effects of audible sound in foam and bubbles are opposite to those found using ultrasound.

In beer, color is defined due to the Maillard reaction that occurs when malted barley is exposed to high temperatures during the kilning process [38]. The yellow color is produced by melanoidins before turning brown. However, a yellow color is maintained when the kilning process is mild. During fermentation, the pH drops and produces a lighter color [35]. Results showed the control had a slightly lighter color (L) and was significantly higher in yellow color (b) than SWF and SWC. However, although non-significant, pH tended to be lower for the control (Table 3). Further studies are needed to assess the cause of the difference in yellow color when applying audible sound to beer processing.

4.2. Descriptive Sensory Evaluation

The trained sensory panel was able to perceive significant differences in FHeight between the three treatments (Figure 4). Similar to the data obtained using the RoboBEER for MaxVol, SWC was rated as the highest in foam height followed by SWF and control. Significant differences were also found for FText, which had SWC rated as the highest and the control as the lowest. The significant differences found using the trained panel in MVisc coincide with those found using the viscometer with SWF rated as the highest in viscosity. This shows that the panel had an appropriate training since they were able to detect differences similar to the more objective measurements. There were no significant differences found in any of the tastes, aromas, and flavors, which was desired since it shows that there was no modification in the sensory profile that could be indicative that there was no breakage of yeast cells with the use of audible sound. Authors such as Martin et al. [39] have found a high yeast cell disruption when applying ultrasound to wine. Therefore, although further research using different types of beer is needed, results from this study showed that using lower frequencies (20–75 Hz) have an effect in foam and bubbles without disturbing the yeast cells, which is a potential treatment that might be applied to beer processing to improve beer quality.

4.3. Validation and Testing of Machine Learning Models

The slightly lower value obtained for SWC from top fermentation and slightly higher value for spontaneous fermentation are mainly due to the higher foam volume and stability, which are two of the main characteristics from spontaneous fermentation beers. However, despite this, the three treatments were classified as top fermentation with a very high accuracy (~99%). This shows that the use of audible sound in beer processing either during fermentation or carbonation stages, even though it improves the beer foamability and bubble size distribution, does not drastically change the nature of the style of beer. Yet, the results obtained from the validation of the ANN model to predict the intensity of sensory descriptors showed that the use of the RoboBEER along with machine learning modeling are potential tools to assess beer quality at the end of the production process to sample every single batch as well as for new products or processing techniques testing in a more objective, effective, and rapid manner when compared to traditional methods.

5. Conclusions

The use of audible sound is a potential treatment to implement in beer processing during the fermentation or carbonation stages to improve the products' quality by increasing the number of small bubbles and increasing foamability and foam stability without modifying the aroma and flavor profile of the specific beer style. However, further research is required to assess the effects of the application of audible sound during both the fermentation and carbonation stages and to find the possible causes for the significant differences in the color. Furthermore, in further research, it would be important to assess the microbiological aspects to analyze yeast viability. The use of the robotic pourer, RoboBEER, the computing vision algorithms and machine learning algorithms are accurate, objective, affordable, and rapid. Tools are used to assess the beer quality in terms of its physical and sensory descriptors for all existing products, new products, and processing techniques.

Supplementary Materials: The following are available online at http://www.mdpi.com/2306-5710/4/3/53/s1, Table S1: Factor loadings of each descriptor in the principal components analysis for the principal components one and two (PC1 and PC2).

Author Contributions: C.G.V. and S.F. contributed equally in the experiment design, data analysis, and paper writing. M.H.L., Y.Q.H., and S.C. contributed in the experiment layout, beer making, and data acquisition. D.T. and F.R.D. contributed to the experiment design and paper writing.

Funding: This research received no external funding.

Acknowledgments: We gratefully acknowledge the support of NVIDIA Corporation with the donation of the Titan Xp GPU used for this research. This research was supported by the Australian Government through the Australian Research Council [Grant number IH120100053] 'Unlocking the Food Value Chain: Australian industry transformation for ASEAN markets.'

Conflicts of Interest: The authors declare no conflict of interest.

References

1. Cromer, A.H.; Vázquez, J.C. *Física Para Las Ciencias de la Vida*; Reverté: Barcelona, Spain, 1981.
2. Serrano Vida, M.; Gil Corral, J. *Musica. Volumen III. Profesores de Educacion Secundaria. Temario Para la Preparacion de Oposiciones*; Ebook; MAD-Eduforma: Sevilla, Spain, 2003.
3. Figura, L.O.; Teixeira, A.A. Acoustical properties. In *Food Physics: Physical Properties—Measurement and Applications*; Springer: Berlin, Germany, 2007.
4. Glasscock, M.E.; Gulya, A.J.; Shambaugh, G.E. *Glasscock-Shambaugh Surgery of the Ear*; BC Decker: Hamilton, ON, Canada, 2003.
5. Vining, W.; Day, R.; Botch, B. *General Chemistry: Atoms First*; Cengage Learning: Boston, MA, USA, 2017.
6. Hassani, S.N.; Bard, R.L. *Real Time Opthalmic Ultrasonography*; Springer: New York, NY, USA, 2012.
7. Malik, H.K.; Singh, A.K. *Engineering Physics*; Tata McGraw Hill Education Private Ltd.: New Delhi, India, 2010.
8. Ferrer, J.F.; Carrera, M.P. *Iniciación a la Física*; Reverté: Barcelona, Spain, 1981.
9. Hassanien, R.H.; Hou, T.-Z.; Li, Y.-F.; Li, B.-M. Advances in effects of sound waves on plants. *J. Integr. Agric.* **2014**, *13*, 335–348. [CrossRef]
10. Czekaj, P.; López, F.; Güell, C. Membrane fouling by turbidity constituents of beer and wine: Characterization and prevention by means of infrasonic pulsing. *J. Food Eng.* **2001**, *49*, 25–36. [CrossRef]
11. Ying, J.C.L.; Dayou, J.; Phin, C.K. Experimental investigation on the effects of audible sound to the growth of Escherichia coli. *Mod. Appl. Sci.* **2009**, *3*. [CrossRef]
12. Jasmine, K.R.; Manohar Das, S.S. Studies on the changes induced by audible sound waves in the total body protein profile of corcyra cephalonica (stainton). *J. Scott Res. Forum* **2008**, *2–4*, 32–39.

13. Aggio, R.B.M.; Obolonkin, V.; Villas-Bôas, S.G. Sonic vibration affects the metabolism of yeast cells growing in liquid culture: A metabolomic study. *Metabolomics* **2012**, *8*, 670–678. [CrossRef]
14. Morales-de la Peña, M.; Welti-Chanes, J.; Martín-Belloso, O. Application of novel processing methods for greater retention of functional compounds in fruit-based beverages. *Beverages* **2016**, *2*, 14. [CrossRef]
15. Abdullah, N.; Chin, N.L. Application of thermosonication treatment in processing and production of high quality and safe-to-drink fruit juices. *Agric. Agric. Sci. Procedia* **2014**, *2*, 320–327. [CrossRef]
16. de Sousa, D.P. *Application of Ultrasounds for Transformation Processes of Agroalimentary Products*; Université d'Avignon: Avignon, France, 2012.
17. Viejo, C.G.; Fuentes, S.; Howell, K.; Torrico, D.; Dunshea, F.R. Robotics and computer vision techniques combined with non-invasive consumer biometrics to assess quality traits from beer foamability using machine learning: A potential for artificial intelligence applications. *Food Control* **2018**, *92*, 72–79. [CrossRef]
18. Viejo, C.G.; Fuentes, S.; Howell, K.; Torrico, D.D.; Dunshea, F.R. Integration of non-invasive biometrics with sensory analysis techniques to assess acceptability of beer by consumers. *Physiol. Behav.* **2018**. [CrossRef]
19. Bamforth, C. Perceptions of beer foam. *J. Inst. Brew.* **2000**, *106*, 229–238. [CrossRef]
20. Donadini, G.; Fumi, M.D.; Faveri, M. How foam appearance influences the Italian consumer's beer perception and preference. *J. Inst. Brew.* **2011**, *117*, 523–533. [CrossRef]
21. Smythe, J.E.; Bamforth, C.W. The path analysis method of eliminating preferred stimuli (PAMEPS) as a means to determine foam preferences for lagers in European judges based upon image assessment. *Food Qual. Preference* **2003**, *14*, 567–572. [CrossRef]
22. Becker, T.; Mitzscherling, M.; Delgado, A. Ultrasonic velocity—A noninvasive method for the determination of density during beer fermentation. *Eng. Life Sci.* **2001**, *1*, 61–67. [CrossRef]
23. Hoggan, J. Ultrasonic hop extraction. *Ultrasonics* **1968**, *6*, 217–219. [CrossRef]
24. Choi, E.J.; Ahn, H.; Kim, M.; Han, H.; Kim, W.J. Effect of ultrasonication on fermentation kinetics of beer using six-row barley cultivated in Korea. *J. Inst. Brew.* **2015**, *121*, 510–517. [CrossRef]
25. Flint, E.B.; Suslick, K.S. The Temperature of cavitation. *Science* **1991**, *253*, 1397–1399. [CrossRef] [PubMed]
26. Galuszka, J. Universities Receive $1 Million to Study Impact of Sound on Fermenting Beer. Available online: https://www.stuff.co.nz/national/education/97068212/universities-receive-1-million-to-study-impact-of-sound-on-fermenting-beer (accessed on 26 June 2018).
27. Fowle, Z. Good Vibrations: Brewers Use Music to Ferment Their Beers. Available online: http://draftmag.com/brewers-ferment-beer-with-music/ (accessed on 26 June 2018).
28. Boxall, A. Does Music Affect the Taste of Beer? Try B&O Play's Beobrew to Find Out. Available online: https://www.digitaltrends.com/home-theater/beoplay-beobrew-music-infused-beer-news/ (accessed on 4 July 2018).
29. Gonzalez Viejo, C.; Fuentes, S.; Li, G.; Collmann, R.; Condé, B.; Torrico, D. Development of a robotic pourer constructed with ubiquitous materials, open hardware and sensors to assess beer foam quality using computer vision and pattern recognition algorithms: RoboBEER. *Food Res. Int.* **2016**, *89*, 504–513. [CrossRef] [PubMed]
30. Gonzalez Viejo, C.; Fuentes, S.; Torrico, D.D.; Howell, K.; Dunshea, F.R. Assessment of beer quality based on a robotic pourer, computer vision, and machine learning algorithms using commercial beers. *J. Food Sci.* **2018**, *83*, 1381–1388. [CrossRef] [PubMed]
31. Wu, H.-Y.; Rubinstein, M.; Shih, E.; Guttag, J.; Durand, F.; Freeman, W. Eulerian video magnification for revealing subtle changes in the world. *Video Magnif.* **2012**. [CrossRef]
32. Torrico, D.D.; Fuentes, S.; Gonzalez Viejo, C.; Ashman, H.; Gunaratne, N.M.; Gunaratne, T.M.; Dunshea, F.R. Images and chocolate stimuli affect physiological and affective responses of consumers: A cross-cultural study. *Food Qual. Preference* **2018**, *65*, 60–71. [CrossRef]
33. Torrico, D.D.; Fuentes, S.; Viejo, C.G.; Ashman, H.; Gurr, P.A.; Dunshea, F.R. Analysis of thermochromic label elements and colour transitions using sensory acceptability and eye tracking techniques. *LWT Food Sci. Technol.* **2018**, *89*, 475–481. [CrossRef]
34. Bamforth, C.; Russell, I.; Stewart, G. *Beer: A Quality Perspective*; Elsevier Science: Cambridge, MA, USA, 2011.
35. Badui Dergal, S.; Cejudo Gómez, H.R.T. *Química de los Alimentos*; Pearson Educación: Naucalpan de Juárez, Estado de Mexico, Mexico, 2006.
36. Delcour, J.A.; Hoseney, R.C. *Principles of Cereal Science and Technology*; AACC International: St. Paul, MN, USA, 2010.

37. Gallego-Juárez, J.; Rodríguez, G.; Riera, E.; Cardoni, A. Ultrasonic defoaming and debubbling in food processing and other applications. In *Power Ultrasonics*; Elsevier: New York, NY, USA, 2015; pp. 793–814. [CrossRef]
38. Piggott, J. *Alcoholic Beverages: Sensory Evaluation and Consumer Research*; Elsevier: Philadelphia, PA, USA, 2011.
39. Martín, J.F.G.; Guillemet, L.; Feng, C.; Sun, D.-W. Cell viability and proteins release during ultrasound-assisted yeast lysis of light lees in model wine. *Food Chem.* **2013**, *141*, 934–939. [CrossRef] [PubMed]

Article

Improving Fermentation Rate during Use of Corn Grits in Beverage Alcohol Production

Deepak Kumar [1], Anna-Sophie Hager [2], Alberto Sun [2], Winok Debyser [2], Bruno Javier Guagliano [2] and Vijay Singh [1,*]

[1] Agricultural and Biological Engineering, University of Illinois at Urbana-Champaign, Urbana, IL 61801, USA; kumard@illinois.edu
[2] Anheuser-Busch InBev nv/sa, Brouwerijplein 1, 3000 Leuven, Belgium; Sophie.Hager@AB-Inbev.com (A.-S.H.); Alberto.Sun@AB-Inbev.com (A.S.); Winok.Debyser@ab-inbev.com (W.D.); brunoguagliano@hotmail.com (B.J.G.)
* Correspondence: vsingh@illinois.edu; Tel.: +1-217-333-9510

Received: 12 December 2018; Accepted: 29 December 2018; Published: 11 January 2019

Abstract: Corn grits are commonly used adjuncts in the brewing industry in the United States, especially for lager beers. The major challenge of using a high amount of adjuncts in the brewing process is reduced levels of nutrients available to yeast during fermentation, which negatively affects the growth and functioning of yeast, and results in sluggish fermentation. The problem is usually addressed by adding external nutrition. The objective of this work was to assess the suitability of corn components other than brewer's grits to improve the fermentation rates. Water obtained after soaking of corn germ, a vital source of lipids and soluble proteins, was investigated as a source of nutrient during brewing of 40:60 (w/w) corn grits and malt mixture. Performance of water-soluble nutrients from germ of two corn verities, yellow dent corn and flint corn, was investigated. Germ soak water was added during corn grits slurry formation before mashing. The addition of germ water increased the free amino nitrogen levels by 37% and Zn concentrations by 3.6 times in the wort, which resulted in up to a 28% higher fermentation rate (between 48 to 72 h of fermentation) and shortened the fermentation time from 120 to 96 h. The use of water obtained from the soaking of flint corn germ resulted in a similar shortening of fermentation time. In another approach, nutrient-rich concentrated germ soak water was directly added into the wort, which also resulted in similar improvements in the fermentation rate as those from adding germ soak water during slurry formation. Due to leaching of micronutrients and soluble proteins, the oil concentrations in the germ increased by more than 30%, enhancing its economic value.

Keywords: beer; adjuncts; fermentation rate; germ; nutrient; FAN

1. Introduction

Beer is the oldest and the most popular alcoholic beverage in the world. Since ancient times, barley has been the most common raw material used for beer brewing. The use of barley provides advantages of easy germination, high starch content, moderate protein content, high amylolytic activity, and wort filtration assistance from husk [1,2]. However, due to increasing demands of sensory modification and specialty beers, and cost optimization, brewing industries are increasing the use of locally available less expensive unmalted grains, known as adjunct grains, in the beer brewing process [3–5]. The use of locally available grains as adjuncts in the brewing process reduces the need for importing malt, provides tax benefits (avoidance of malt tax in some countries), reduces the carbon footprint, and supports local farmers [6]. Corn, wheat, rice, unmalted barley, oats, and sorghum are some of the major adjunct grains used. High amounts of adjuncts are used to produce specialty beers. A type of beer known as "happoshu" is produced in Japan that contains more than 75% rice as adjuncts [6]. The use of adjuncts also provides advantages of increased foam stability, various color and flavor options, and health benefits (e.g., reduced gluten levels). The potential cost savings are highly dependent on the type of adjunct used, local prices of adjunct grains, malt availability, and other costs of productions [4]. Although the benefits of adjuncts have been realized and highlighted in the last few years only, the use of adjuncts is as old as beer itself. For example, the Sumerians (Mesopotamia; Lady Pu-Abi Queen of Ur (2600BC)) had a beer that was made from barley and emmer (ancient wheat). Raw wheat has been used as an adjunct in the production of lambic (a type of spontaneous fermented beer), in Belgium for centuries. Currently, 85% to 90% of all of the beer in the world is produced with adjuncts used in the process [5].

Due to large availability, corn grits are the most commonly used adjuncts in the United States, especially for lager beers. The use of 30% corn grits as adjuncts can reduce brewing costs by about 8% [4]. In addition to ready availability and lower cost, the use of corn adjuncts provides benefits in the form of color adjustments and sweet and fuller flavor profile. The profile of fermentable sugars and dextrins from enzymatic conversion corn-based adjuncts is similar to that of malt [2]. In 2017, about 150 million bushels of corn were used in beverage alcohol production in the United States [7].

The use of adjuncts in brewing also has several disadvantages, including the need of a cereal cooker, low enzymatic power, and reduced levels of nutrients for yeast. A cereal cooker is needed for starch gelatinization and commercial enzymes are often used to overcome the limitation of lower amylolytic activity [4]. The major limitation that use of adjuncts in the brewing process leads to is the reduced levels of nutrients available to yeast during fermentation. Most of the adjuncts used in the industry have relatively low protein levels, and replacement of malt with these adjuncts decreases the concentrations of nitrogen-containing substances during fermentation, also known as protein dilution in wort [5]. Nitrogen is an essential element required for yeast growth and synthesis of cellular proteins and other cell compounds. Individual amino acids, small peptides, and some ammonium ions produced from the proteolysis of proteins in barley malt are the main nitrogen sources for yeast metabolism in wort [8]. Free amino nitrogen (FAN) level, collectively accounting for individual amino acids and small peptides (dipeptides and tripeptides), are commonly monitored to ensure efficient yeast cell growth in order to achieve a desirable fermentation performance [8]. Poreda et al. (2013) reported that even with similar protein content in malt and corn grits, FAN concentration in wort obtained with the addition of 20% grits was 7% lower (223 ppm vs. 240 ppm) compared to pure malt wort [4]. It has been reported that proteins in adjuncts do not get hydrolyzed as efficiently as malt protein and result in a deficiency of total soluble nitrogen in wort [4,9]. Partially hydrolyzed proteins remain insoluble and are separated along with spent grains during filtration [5]. Other than reduced free nitrogen levels, the use of high amounts of adjuncts also decreases the concentrations of other nutrients, such as inorganic ions and vitamins, which required for efficient yeast growth [10]. The nitrogen limitation and deficiency of micronutrients results in poor yeast viability, which leads to retarded or sluggish fermentation [5,11]. To overcome this issue, the wort obtained during the use of high percentage of adjuncts is usually supplemented with external nutrients in order to achieve an efficient fermentation. The challenge of inefficient fermentation due to lack of yeast nutrition has also been reported in the form of the use of dry-fractionated corn for fuel ethanol production [12,13]. During whole corn fermentation, germ is a vital source of lipids and water-soluble proteins, which help in maintaining the membrane integrity and yeast performance. However, flaking grits and brewer grits, obtained from the dry-fractionation of corn, do not contain the germ fraction and have a relatively small amount of protein, oil, and other micronutrients, compared to whole corn, which causes nutrient deficiency to yeast during fermentation. The addition of yeast extract, lipid supplementation, and leaving some germ (minimum 20%) behind, are some successful reported approaches to improving fermentation of dry-fractionated corn during fuel ethanol production [12–14]. However, due to concerns of the negative effects of lipids on beer flavor and foam, these approaches cannot be applied in the beverage alcohol production process. One approach that could have potential application in the brewing process is the extraction of soluble nutrients from corn germ by soaking and the use of that water in the process.

The water would contain only soluble proteins and micronutrients but not lipids. Murthy et al. (2006) reported that the use of germ soak water (2 h soaking) in the dry grind process resulted in higher final ethanol concentrations (14.7% vs. 12.3% (v/v)) with a relatively small amount of residual glucose [14]. Juneja et al. (2018) reported that the use of water from germ soaked for a longer time (12 h instead of 2 h) can result in complete fermentation (no residual glucose) and achieve a fermentation profile similar to that obtained with the addition of excess B vitamins or protease enzymes [15]. Both of these studies only reported the fermentation improvements but did not determine the actual nutrition enhancement with the addition of germ soak water.

The objective of this study was to investigate the feasibility of using corn germ soak water to improve yeast nutrition and fermentation rates in the brewing process using 40% corn grits adjuncts. Performance of water-soluble nutrients from germ of two corn verities, yellow dent corn and flint corn, was investigated. In addition to a new process development, individual amino acids, total FAN levels, and other important nutrients (e.g., Zn concentrations) in the germ water and wort were determined to enhance the understanding of the potential benefits of implementing this approach.

2. Materials and Methods

2.1. Materials

Degermed yellow brewing grits and corn germ (from yellow dent corn) samples were obtained from a commercial corn dry milling plant (Bunge, Danville, IL, USA). Malt and flint corn samples were obtained from a commercial brewing company (Anheuser-Busch InBev, Brouwerijplein 1, 3000 Leuven, Belgium). Germ from the flint corn was separated using a 1 kg laboratory scale protocol outlined by Rausch et al. (2009) [16]. All materials were stored at 4 °C in refrigerator until use.

Active dry brewing yeast Saflager S-189 yeast, used for the fermentation in the current study, was obtained from the Fermentis-Lesaffre Yeast Corporation (Milwaukee, WI, USA). Novozymes (Bagsvaerd, Denmark) generously donated Termamyl 120 L, Type L, a thermostable bacterial α-amylase preparation.

2.2. Corn Grits and Germ Composition

Starch content in the ground corn grits was determined by the modified acid hydrolysis method [17] and was found to be 79.64%. About 1 g of ground corn grits were mixed with 50 mL of 0.4N HCl in 100-mL autoclavable glass bottles. After mixing, the slurry was autoclaved for 1 h at 126 °C in a laboratory scale autoclave (Napco Model 9000D, Thermo Fisher 157 Scientific, Waltham, MA, USA). Glucose recovery factors were determined using pure glucose and starch samples. After cooling, 1-mL aliquot samples were withdrawn and centrifuged at $1500 \times g$ for 5 min (Model 5415 D, Brinkmann–Eppendorf, Hamburg, Germany). The supernatants were analyzed in the HPLC for glucose concentration determination.

Germ (before and after soaking) samples were analyzed for oil (AOAC method 920.39) content by a commercial analytical laboratory (Illinois crop improvement association, Champaign, IL, USA) [18]. All analyses were conducted in duplicates.

2.3. Brewing Process

A simple schematic of the lab-scale process used for the beverage alcohol production is shown in Figure 1. Corn grits and malt samples were ground in a hammer mill (model MHM4, Glen Mills, 126 Clifton, NJ, USA) equipped with a 1.0-mm sieve. The moisture contents of the ground samples were determined by drying flour at 135 °C for 2 h (Approved Method 44-19, AACC International 2000).

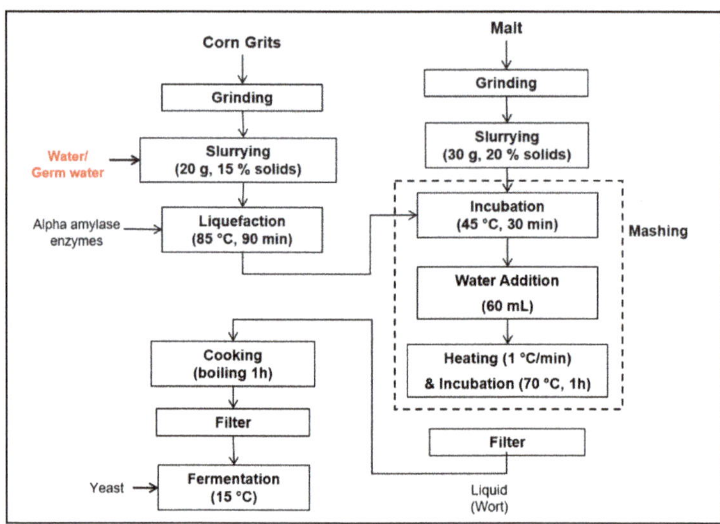

Figure 1. Schematic of steps followed for fermentation of corn grits and malt.

Malt and corn grits were used in 60:40 ratios on a mass basis. Twenty grams (dry weight) of ground corn grits was mixed with a calculated amount of deionized water (for control) or germ soak water (for treatments) to form a slurry with 15% solids, and the pH of the slurry was adjusted to 5.8 (recommended optimum pH for the α-amylase). Germ soak water was obtained by soaking the germ in deionized water (1:10, germ:water on a weight basis) at 52 °C for 12 h, with continuous shaking at 125 rpm. After soaking, the liquid was vacuum-filtered through Whatman No. 4 filter paper and used to make corn slurry. The corn grits starch was liquefied by adding 10 µL of α-amylase and agitating the slurry at 85 °C for 120 min using an infrared heated reactor system equipped with mini-stainless steel tubular reactors that are locked onto a rotating carrousel (Labomat BFA-12, Werner Mathis AG, Switzerland). Simultaneously, 30 g (dry weight) of ground malt was mixed with a calculated amount of deionized water to form a mixture with 20% solids, and incubated at 45 °C for 30 min, with continuous shaking at 125 rpm. After incubation, malt slurry was mixed with the liquefied corn slurry (cooled to 45 °C), and 60 mL of preheated water (45 °C) was added in the mixture. The mixture was heated to 70 °C at the heating rate of 1 °C per minute, and incubated for 60 min in the mini-stainless steel tubular reactors using a Labomat incubator. The mash was cooled down to room temperature and filtered using Whatman No. 4 filter paper to separate liquid (wort) and solids (spent grains). The filtered wort was boiled on a heating plate for about 1 h and filtered again (using Whatman No. 4 filter paper) after cooling. Yeast inoculum was prepared by sprinkling dry yeast in ten times the amount of water and incubating at 23 °C for 20 min, followed by gentle stirring for 30 min. The yeast was inoculated in the wort at the dosage of 100 g/hL, as recommended by the supplier. The fermentation was performed at 15 °C for 144 h with continuous agitating at 125 rpm. About 2 mL of samples were withdrawn every 24 h to monitor the fermentation.

2.4. Sample Analysis and Ethanol Production Rate

Samples collected during the fermentation process were analyzed for sugars and ethanol concentrations using HPLC, equipped with an ion-exclusion column (Aminex HPX-87H, Bio-Rad, Hercules, CA, USA). The mobile phase used was 50 mM sulfuric acid, eluted at 50 °C and 0.6 mL/min. The amounts of sugars and ethanol were quantified using a refractive index detector and calibration based on pure sugars and ethanol standards.

Ethanol production rates (fermentation rates, %v/v-h) between different time points were calculated using the following equation.

$$\text{Ethanol production rate} = \frac{E_{t2} - E_{t1}}{t_2 - t_1} \tag{1}$$

where E_{t2} and E_{t1} are ethanol concentrations (%v/v) at fermentation times t_2 and t_1 in hours.

2.5. Total Free Amino Nitrogen and Amino Acid Profile

Total free amino nitrogen (FAN) concentrations in various germ soak water samples were determined using the ninhydrin colorimetric method (Official Method 945.30L, AOAC 1980) [19]. Quantification of individual amino acids concentration in germ soak water and wort was conducted at the University of Missouri Agricultural Experiment Station Chemical Laboratories (ESCL) using the AOAC official methods (Method 982.30 E(a,b)) [20].

2.6. Zinc Determination

Zn concentrations in the germ water samples were determined using ICP (Inductively Coupled Plasma) analysis, performed in the Microanalysis Laboratory, School of Chemical Sciences at the University of Illinois.

3. Results and Discussion

3.1. Effect of Germ Water Addition

The wort obtained from the processing of a mixture of malt and corn grits (60:40, weight basis) as described in Figure 1, contained 1.1% glucose, 8% maltose, 1.5% maltotriose, and 0.25% fructose. The sugar concentrations were similar in wort obtained from control samples (no germ water addition) and wort from germ water supplemented mash. The sugar profile and concentrations from the current process match closely with the values reported by Stewart et al. (2013) for 30% corn adjunct wort [10]. Figure 2 illustrates the comparison of ethanol concentration during fermentation, for control and germ water addition. Use of germ soak water improved the fermentation profile significantly compared to that of the control. Yeast preferred glucose over maltose during fermentation, and almost all of the glucose was consumed in the first 48 h for both cases (Figure 3). Fermentation profiles of control and germ soak water samples were almost similar in the first 48 h (until the presence of glucose). However, maltose to ethanol conversion rates were higher for samples with the addition of germ soak water. Almost all of the maltose was converted to ethanol in 96 h for the germ soak water samples, compared to 1.47% (w/v) residual maltose for the control. The ethanol concentration of 6.32% (v/v) was 18.5% higher compared to the control (5.33% (v/v)) after 96 h of fermentation. Ethanol production rates for the germ water treatment were 16% and 28% higher than that of control between 24–48 and 48–72 h of fermentation, respectively. Due to low sugar availability, fermentation rate drops after 72 h and it was lower for the germ water treatment compared to the control after 96 h of fermentation. The higher fermentation rate could be attributed to the improved functioning of the yeast due to the availability of free amino acids and other nutrients present in germ soak water.

Total FAN concentrations found in the water obtained from germ soaked in various amounts of water (germ:water weight ratios of 1:10, 1:5, 1:2.5) are presented in Table 1. As discussed earlier, the brewing industry uses FAN concentration as an indicator of available assimilable nitrogen. Free amino nitrogen is essential for the synthesis of structural proteins and enzymatic proteins required for yeast growth and metabolic reactions [5]. Adequate levels of FAN are critical for the yeast to perform efficient fermentation [8,21]. FAN is considered an index for predicting yeast viability, vitality, and fermentation efficiency [10]. Water obtained from 1:10 germ:water (weight basis) ratio, used in this experiment, contained 127 ppm of FAN. The use of this water to make corn mash resulted in total FAN of 170 ppm (11.5 mg/L/Plato) in the wort, compared to 125 ppm (8.4 mg/L/Plato) in wort from the control. Heldt-Hansen et al. (2011) recommended a FAN concentration of 9–14 mg/L/Plato and 10–18 mg/L/Plato in unmalted barley (adjuncts) wort and all-malt worts, respectively, to achieve comparable fermentation performance [5]. The FAN level in the wort with the use of germ soak water is within the recommended range. The FAN concentrations in the germ water increased with a decrease in water used while soaking. As the amount of soaking water was reduced to half (1:5 germ:water ratio), the FAN concentrations almost doubled. Similarly, for a 1:2.5 germ:water ratio, the FAN concentration was more than 4 times that of water compared to the treatment with 1:10 germ:water soaking (Table 1). The effect of shaking during the germ soaking on FAN concentrations was also investigated. No significant difference in FAN concentrations in the water obtained from germ soaked with or without shaking was observed.

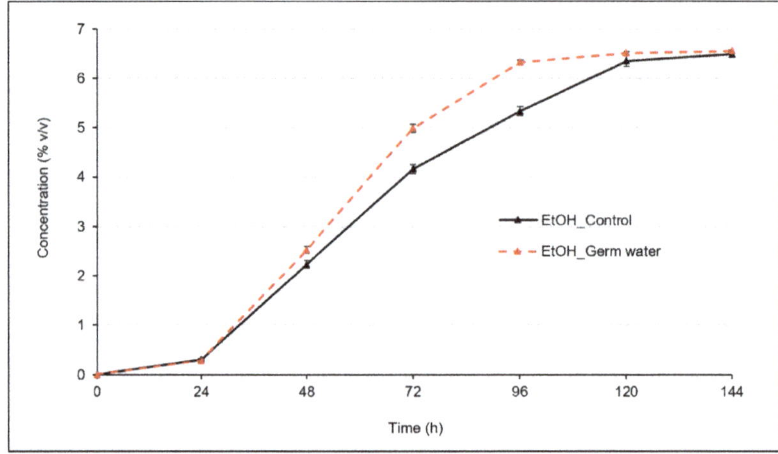

Figure 2. Ethanol concentration during fermentation without (control) and with addition of germ soak water.

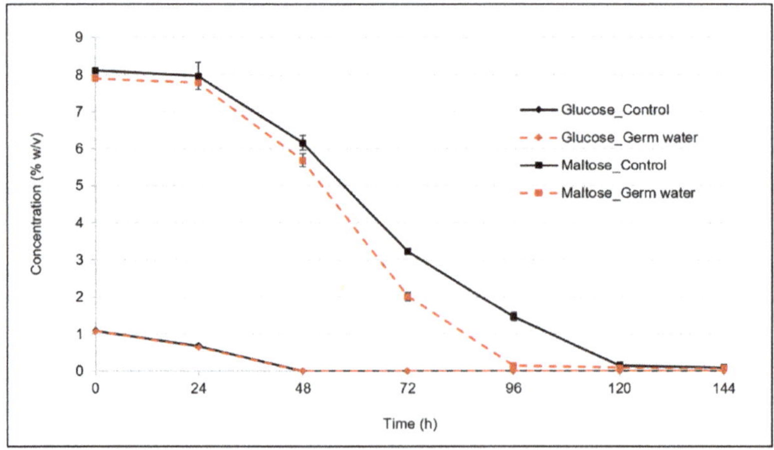

Figure 3. Glucose and maltose concentration during fermentation without (control) and with addition of germ soak water.

Table 1. Free amino nitrogen (FAN) concentrations in the water obtained from germ soaked under various conditions.

Soaking (Germ:Water by Weight)	Concentration (ppm)
1:10	127 ± 8
1:5	250 ± 12
1:2.5	517 ± 12

Other than reducing overall free nitrogen levels, the use of adjuncts is believed to alter the proportion of the various amino acids in the wort, which is also important for yeast functioning [5]. Figure 4a illustrates the concentration of amino acids in the germ water. Correspondingly, the amino acid concentrations for wort from control and germ water supplemented samples are provided in Figure 4b. Total amino acid concentrations in germ water, wort from control, and wort from germ water supplemented samples were found to be 100 ppm, 115 ppm, and 146 ppm, respectively. The concentrations of lysine and leucine, amino acids categorized in Group A (fast absorption) and Group B (intermediate absorption), were double for wort in germ water supplemented samples compared to wort from control samples. Other than these, the amounts of phenylalanine and hydroxylysine were also double in wort from germ water supplemented samples.

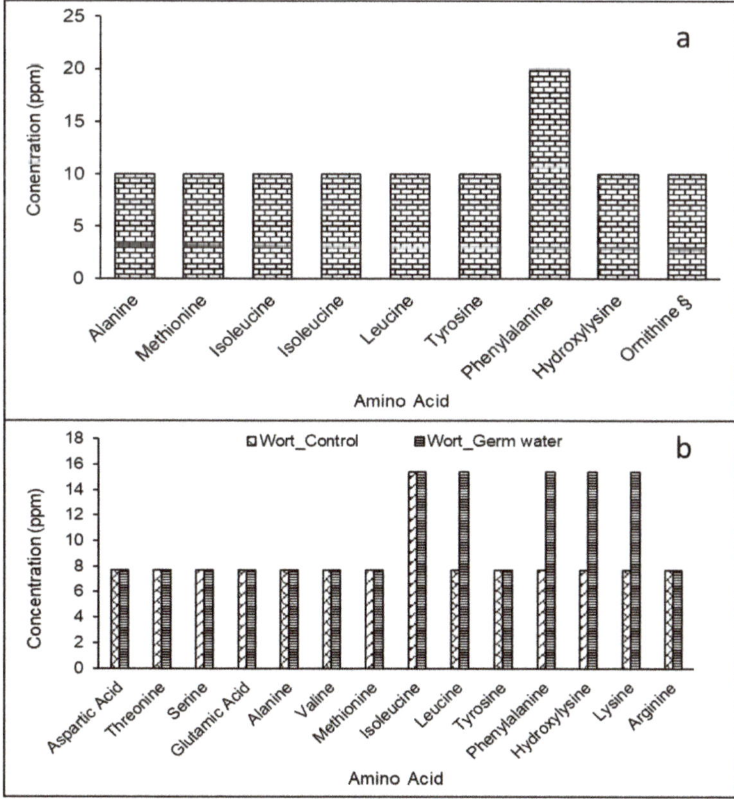

Figure 4. Amino acid concentrations in (**a**) liquid obtained after germ soaking and (**b**) wort from wort obtained from control samples and from germ water supplemented samples. Data in Figure 4a is average of two replications.

Other than FAN, some trace metals—such as Zn, which is considered to be an important trace metal that plays an essential structural and functional role in yeast cells—act as a modulator of environmental stress and are important for the fermentation process [22,23]. Although Zn concentrations are at a sub-ppm level in the wort, even these small amounts are critical for efficient fermentation of wort [24]. Zinc concentrations below 0.1 ppm can lead to fermentation problems, especially decreased specific rate of fermentation [24,25]. The germ water used in this experiment (from 1:10 germ:water ratio) contained 1.49 ppm of Zn. The use of this water in the brewing process resulted in a Zn concentration of 0.33 ppm in the wort, compared to 0.09 ppm in wort from the control. The adequate levels of Zn in the wort from the germ water supplemented process can be another factor for the observed increased fermentation rates.

3.2. Effect of Germ Water Addition from Flint Corn Germ

For real application of this approach (using water-soluble nutrients from corn germ) in the brewing industry, it is critical to ensure the availability of the nutrient source. As the types of corn used vary among industries, it was important to investigate the performance of germ obtained from various corn types. The effect of water obtained from the soaking of germ from the fractionation of flint corn was also investigated on the fermentation performance of lager yeast. Germ was separated from the flint corn milled employing a slightly adapted laboratory-scale (1 kg) dry-milling procedure, outlined by Rausch et al. (2009). About 40 to 50 g of germ was obtained from the milling of 1 kg of corn. Germ water collection and fermentation procedures were similar to that performed with the yellow dent corn germ. Sugar consumption and ethanol production profiles during the fermentation are presented in Figure 5. Similar to the case of yellow dent corn germ, the use of water from the soaking of flint corn germ improved the fermentation of maltose significantly compared to that of the control. The fermentation profiles of the control and germ soak water samples were almost similar in the first 48 h, until all of the glucose was consumed. The ethanol production rate for the germ water treatment was 16% and 42% higher than the control between 24–48 and 48–72 h of fermentation, respectively. Almost all of the maltose and maltotriose were consumed in 96 h of fermentation in the case of germ soak water treatment contrary to 1.31% and 0.38% (w/v) residual maltose and maltotriose, in the case of control. Due to low sugar availability, the fermentation rate dropped and was lower than the control after 72 h and it was almost zero after 96 h of fermentation. Although all the maltose was consumed in 96 h in both cases, the increase in fermentation rate between 48–72 h in the current case was compared to that observed with the addition of water obtained from yellow dent corn germ (42% vs. 28% increase). This could be due to the availability of relatively higher free nitrogen. Water obtained from the soaking of flint corn germ at 1:10 germ:water (weight basis) ratio contained 195 ppm of FAN, which was significantly higher than the water from dent corn germ (127 ppm). The total FAN concentration in wort with the use of flint corn germ was 193 ppm (13.1 mg/L/Plato), compared to 128 ppm (8.7 mg/L/Plato) in the control of this experiment, and 170 ppm (11.5 mg/L/Plato) in wort from the addition of yellow dent corn germ. The relatively higher amount of FAN in the water from flint corn germ compared to dent corn germ could be due to genetic differences between two corn verities and different quality germ obtained from laboratory scale milling (for flint corn) and commercial milling (for dent corn). The high quality of the flint corn germ was also indicated by higher oil (28.8 vs. 21.6%, dry basis) and protein (17.3% vs. 15.5%, dry basis) contents in the flint corn germ compared to the yellow dent corn germ.

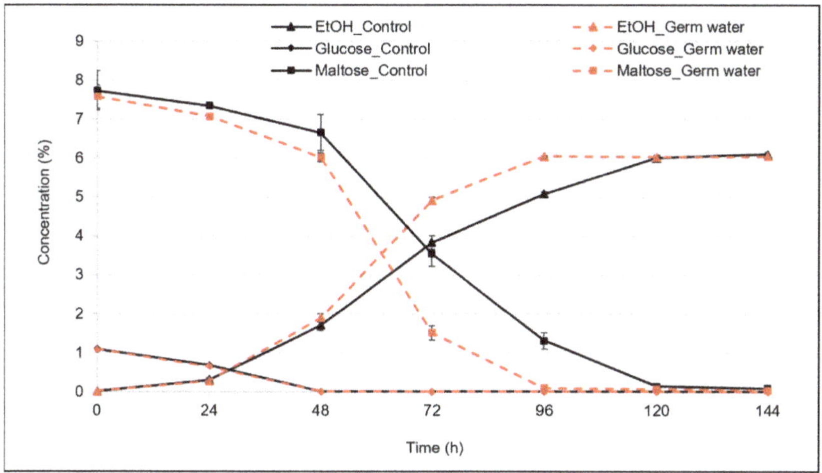

Figure 5. Ethanol (% v/v), glucose (% w/v), and maltose (% w/v) concentrations during fermentation without (control) and with the addition of flint corn germ soak water.

3.3. Use of Concentrated Germ Water in Wort

In the previous experiments, germ water was added during the slurry formation before mashing (Figure 1). To explore the possibility of using these water-soluble nutrients without making many changes in the currently used brewing process, another approach of adding germ water directly in the wort was also investigated. However, the addition of germ water in wort would result in dilution. To provide the same amount of nutrients (as in previous experiments), the addition of germ water (1:10 germ to water ratio soaking) would result in about 40% dilution. This problem was addressed by using germ soak water containing a higher concentration of nutrients. The total FAN in the water obtained from 1:2.5 germ to water ratio (referred as 4X water in the manuscript) was a little more than 4 times the FAN concentration in water from 1:10 germ to water ratio (referred as 1X water in the manuscript) during soaking (Table 1). Due to the high amount of FAN, a relatively small amount of this water would be required to be added to the wort, which would reduce the dilution (only 10%). Further concentration (using less than 2.5 times water) was not possible due to insufficient water available to soak all of the germ. Germ (from yellow dent corn) was soaked in water (1:2.5 germ to water ratio, mass basis) and incubated at 52 °C for 12 h with continuous shaking at 125 rpm. In this experiment, 27 mL of this water was mixed in 265 mL wort, before the boiling step. The rest of the steps were similar to the previous process, as explained in Figure 1. In addition to the current treatment (adding 4X water in wort) and control (no germ water), another treatment (1X water) following the original process (adding germ water from 1:10 germ to water ratio to make corn mash) was also conducted for comparison purposes. To keep the wort concentrations similar in all three cases, 27 mL of deionized water was added in the wort before boiling for control and 1X water. All treatments were conducted in replication.

Figure 6 illustrates the ethanol production rate comparison for three cases: (i) control, (ii) use of germ soak water to make corn mash (1X water), (iii) use of concentrated (4X) germ soak water in the wort. The addition of concentrated germ soak water in wort increased the fermentation rate compared to the control. Both treatments (the addition of 4X water in wort or making corn slurry with germ water) were equally efficient in increasing the fermentation speed. The ethanol concentration was almost the same for both treatments at all the time points. Less than 0.1% (w/v) maltose was observed after 96 h of fermentation, compared to 0.73% (w/v) for the control, for which the maltose concentration was found to be 0.1% after 120 h of fermentation. The fermentation rate was 30% higher than the control between 48 and 72 h of fermentation. These results indicate that germ water can be concentrated and added to the wort to fasten the fermentation.

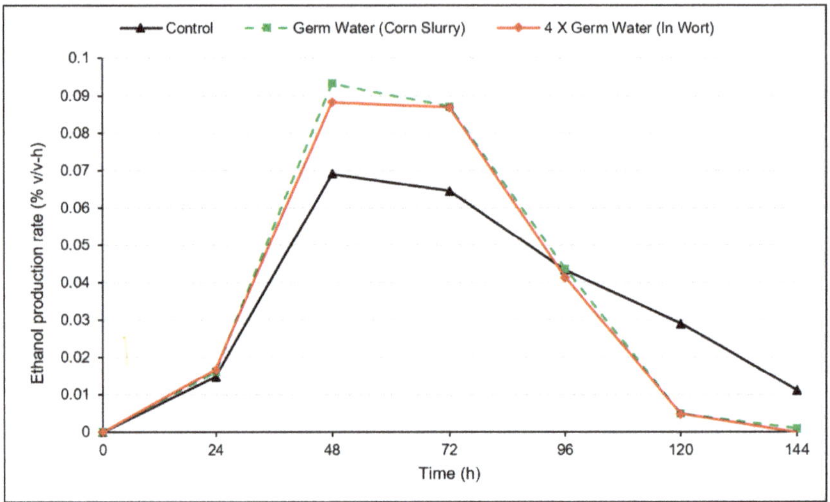

Figure 6. Rate of ethanol production during fermentation for three cases: (i) control, (ii) use of germ soak water in corn mash, (iii) use of concentrated germ soak water in wort. Data in the graphs is the average of two replications.

3.4. Change in Composition of Germ

Corn germ is valuable for its oil content, which directly decides its market value [26]. The extraction of soluble protein and other micronutrients during soaking would increase the oil content in germ and improve its economic value. The oil content of soaked dent corn germ increased by 33% (21.6% to 28.7%, dry basis) compared to raw germ. The increase in oil content of flint corn germ was found to be 29%. Due to increased oil content, the germ would have a higher market price, which could compensate the cost required for soaking and filtration to make germ soak water. The protein content in soaked dent corn germ (14.7%) was 5% less compared to raw germ (15.5%). This reduction in protein content further indicates that soluble proteins were extracted during germ soaking, which enhanced the yeast functioning and fermentation speed.

4. Conclusions

Due to low protein content, the use of high amounts of corn grits as adjuncts causes nutrient deficiency to yeast during fermentation and leads to sluggish fermentation. This work developed economical process strategies to improve the fermentation rate by providing these nutrients from cereal-based origin. Water obtained from the soaking of corn germ was used as a nutrient source to improve yeast performance during the brewing process using 40% corn grits as adjuncts. The addition of germ water improved the FAN levels in the wort, which enhanced yeast functioning and shortened the fermentation time from 120 to 96 h. The water obtained from flint corn germ had higher amounts of FAN compared to dent corn germ, however, the total maltose consumption time was 96 h in both cases. The study also demonstrated that the nutrients can be concentrated up to 4 times by proportionally decreasing the degree of dilution during the soaking process, and the concentrated water can be added directly to the wort to improve fermentation speed. Soaking of germ increased oil content by 33% in the germ, improving its market value.

Author Contributions: D.K. performed experiments, analyzed data, and wrote the manuscript. D.K., V.S., and A.-S.H. designed the study and reviewed results. B.J.G. helped with data analysis and the design of experiments. A.-S.H. and W.D. reviewed and edited the manuscript. All authors read and approved the manuscript.

Funding: Partial funding for this study was provided by Anheuser Busch InBev.

Conflicts of Interest: Some of the coauthors (Anna-Sophie Hager, Alberto Sun, Winok Debyser, Bruno Guagliano) are current and former employees of Anheuser Busch InBev. However, in this study a new process was developed and no Anheuser Bush InBev products or processes were evaluated or compared with any other company's products.

References

1. Li, Q.; Wang, J.; Liu, C. Beers. In *Current Developments in Biotechnology and Bioengineering*; Elsevier: New York, NY, USA, 2017; pp. 305–351. Available online: https://www.sciencedirect.com/book/9780444636669/current-developments-in-biotechnology-and-bioengineering (accessed on 11 January 2019).

2. Bravi, E.; Sensidoni, M.; Floridi, S.; Perretti, G. *Fatty Acids Composition Differences between Beers Made With All-Malt and Brewer's Corn Grits and Malt*; Technical quarterly; Food and Agriculture Organization (FAO): Rome, Italy, 2009.

3. Zhuang, S.; Shetty, R.; Hansen, M.; Fromberg, A.; Hansen, P.B.; Hobley, T.J. Brewing with 100% Unmalted Grains: Barley, Wheat, Oat and Rye. *Eur. Food Res. Technol.* **2017**, *243*, 447–454. [CrossRef]

4. Poreda, A.; Czarnik, A.; Zdaniewicz, M.; Jakubowski, M.; Antkiewicz, P. Corn grist adjunct–application and influence on the brewing process and beer quality. *J. Inst. Brew.* **2014**, *120*, 77–81. [CrossRef]

5. Bogdan, P.; Kordialik-Bogacka, E. Alternatives to malt in brewing. *Trends Food Sci. Technol.* **2017**, *65*, 1–9. [CrossRef]

6. Cooper, C.; Evans, D.; Yousif, A.; Metz, N.; Koutoulis, A. Comparison of the impact on the performance of small-scale mashing with different proportions of unmalted barley, Ondea Pro®, malt and rice. *J. Inst. Brew.* **2016**, *122*, 218–227. [CrossRef]

7. USDA-NAAS. *Grain Crushings and CoProducts Production 2017 Summary*; United States Department of Agriculture: Washington, DC, USA, 2018.

8. Lekkas, C.; Stewart, G.; Hill, A.; Taidi, B.; Hodgson, J. The importance of free amino nitrogen in wort and beer. *Tech. Q.-Master Brew. Assoc. Am.* **2005**, *42*, 113.

9. Zembold-Gula, A.; Blazewicz, J.; Wojewodzka, K. *Protein Compounds in Brewing Worts Produced with Addition of Naked Barley Grain*; Zywnosc Nauka Technologia Jakosc (Poland); Food and Agriculture Organization (FAO): Rome, Italy, 2008.

10. Stewart, G.G.; Hill, A.E.; Russell, I. 125th anniversary review: Developments in brewing and distilling yeast strains. *J. Inst. Brew.* **2013**, *119*, 202–220. [CrossRef]

11. Yano, M.; Tsuda, H.; Imai, T.; Ogawa, Y.; Ohkochi, M. The effect of barley adjuncts on free amino nitrogen contents in wort. *J. Inst. Brew.* **2008**, *114*, 230–238. [CrossRef]

12. Murthy, G.S.; Singh, V.; Johnston, D.B.; Rausch, K.D.; Tumbleson, M. Improvement in fermentation characteristics of degermed ground corn by lipid supplementation. *J. Ind. Microbiol. Biotechnol.* **2006**, *33*, 655–660. [CrossRef] [PubMed]

13. Ramchandran, D.; Wang, P.; Dien, B.; Liu, W.; Cotta, M.A.; Singh, V. Improvement of Dry-Fractionation Ethanol Fermentation by Partial Germ Supplementation. *Cereal Chem.* **2015**, *92*, 218–223. [CrossRef]

14. Murthy, G.S.; Singh, V.; Johnston, D.B.; Rausch, K.D.; Tumbleson, M. Evaluation and strategies to improve fermentation characteristics of modified dry-grind corn processes. *Cereal Chem.* **2006**, *83*, 455–459. [CrossRef]

15. Juneja, A.; Kumar, D.; Singh, V. Germ soak water as nutrient source to improve fermentation of corn grits from modified corn dry grind process. *Bioresour. Bioprocess.* **2017**, *4*, 38. [CrossRef] [PubMed]

16. Rausch, K.D.; Pruiett, L.E.; Wang, P.; Xu, L.; Belyea, R.L.; Tumbleson, M. Laboratory Measurement of Yield and Composition of Dry-Milled Corn Fractions Using a Shortened, Single-Stage Tempering Procedure. *Cereal Chem.* **2009**, *86*, 434–438. [CrossRef]

17. Vidal, B.C., Jr.; Rausch, K.D.; Tumbleson, M.; Singh, V. Protease treatment to improve ethanol fermentation in modified dry grind corn processes. *Cereal Chem.* **2009**, *86*, 323–328. [CrossRef]

18. AOAC. *Official Methods of the Association of Official Analytical Chemists Washington, DC*; Vol. Methods 920.39, 924.05, 973.18, 990.03; AOAC: Rockville, MD, USA, 2003.

19. Vidal, B.C., Jr.; Johnston, D.B.; Rausch, K.D.; Tumbleson, M.; Singh, V. Germ-derived FAN as nitrogen source for corn endosperm fermentation. *Cereal Chem.* **2011**, *88*, 328–332. [CrossRef]

20. AOAC. *Official Methods of Analysis of AOAC International: Arlington, VA.*; Vol. Method 982.30 E(a, b, c) chp. 45.3.05; AOAC: Rockville, MD, USA, 2006.

21. Lekkas, C.; Stewart, G.; Hill, A.; Taidi, B.; Hodgson, J. Elucidation of the role of nitrogenous wort components in yeast fermentation. *J. Inst. Brew.* **2007**, *113*, 3–8. [CrossRef]

22. Rowell, J.L. *Exploring Fermentation Rate in Beer and Hard Cider Brewing*; Western Carolina University: Cullowhee, NC, USA, 2015.

23. Walker, G.M. Metals in yeast fermentation processes. *Adv. Appl. Microbiol.* **2004**, *54*, 197–230. [PubMed]

24. De Nicola, R.; Walker, G.M. Accumulation and cellular distribution of zinc by brewing yeast. *Enzyme Microb. Technol.* **2009**, *44*, 210–216. [CrossRef]

25. Bromberg, S.K.; Bower, P.A.; Duncombe, G.; Fehring, J.; Gerber, L.; Lau, V.K.; Tata, M. Requirements for zinc, manganese, calcium, and magnesium in wort. *J. Am. Soc. Brew. Chem.* **1997**, *55*, 123–128. [CrossRef]

26. Johnston, D.B.; McAloon, A.J.; Moreau, R.A.; Hicks, K.B.; Singh, V. Composition and economic comparison of germ fractions from modified corn processing technologies. *J. Am. Oil Chem. Soc.* **2005**, *82*, 603–608. [CrossRef]

Article

Why Craft Brewers Should Be Advised to Use Bottle Refermentation to Improve Late-Hopped Beer Stability

Carlos Silva Ferreira, Etienne Bodart and Sonia Collin *

Earth and Life Institute ELIM, Université catholique de Louvain, Croix du Sud, 2 box L7.05.07,
B-1348 Louvain-la-Neuve, Belgium; carlos.silva@uclouvain.be (C.S.F.); etienne.bodart@uclouvain.be (E.B.)
* Correspondence: sonia.collin@uclouvain.be; Tel.: +32-10472913

Received: 28 February 2019; Accepted: 13 May 2019; Published: 4 June 2019

Abstract: The aromatic complexity of craft beers, together with some particular practices (use of small vessels, dry hopping, etc.), can cause more oxidation associated with pre-maturated colloidal instability, Madeira off-flavors, bitterness decrease, and aroma loss. As bottle refermentation is widely used in Belgian craft beers, the aim of the present work is to assess how this practice might impact their flavor. In fresh beers, key flavors were evidenced by four complementary techniques: short-chain fatty acids determination, esters analysis, XAD-2 extract olfactometry, and overall sensory analysis. In almost all of the fresh beers, isovaleric acid was the sole fatty acid found above its sensory threshold. Selected samples were further analyzed through natural aging at 20 °C. The presence of yeast in the bottle minimized the *trans*-2-nonenal released from Schiff bases and proved less deleterious than suggested by previous studies with regard to fatty acid release and ester decrease through aging. Furthermore, according to the yeast species selected, some interesting terpenols and phenols were produced from glucosides during storage.

Keywords: craft beer; bottle refermentation; AEDA; short-chain fatty acids; beer aging

---

1. Introduction

Worldwide, over the last few decades, the production of craft beers has grown significantly, with new commercial products launched onto the market every day. Craftsmen can bring distinctive flavors to their beers by working with special malts, dual-purpose hop varieties (with or without dry hopping), spices, and/or specialty yeasts. These are known to impart fruity esters [1,2] and, in some cases, typical phenolic flavors [3] (e.g., 4-vinylguaiacol brought by phenolic off-flavor (POF(+)) yeasts). In addition, odorous heterocyclic compounds can be issued from colored malts [4,5], while hop terpenols and polyfunctional thiols bring pleasant citrus and exotic flavors to late- and dry-hopped beers [6–8]. Unfortunately, the use of small vessels and craftsmanship, by definition, lead to a higher risk of oxidation and shelf-life decrease.

Beer aging has been the focus of much interest for decades, with the development of worldwide beverage exchanges. The Dalgliesh plot [9] describes aromatic changes occurring in lager beers during storage. A linear decrease in bitterness (degradation of isohumulones and/or humulinones) coincides with an increase in sweet aroma and toffee flavor, together with the well-known cardboard taint (caused by *trans*-2-nonenal) [10–13] and *ribes* off-flavor (a catty smell linked to the presence of 3-sulfanyl-3-methylbutyl formate) [14,15]. Aging of specialty beers is even more complex, with defects such as Madeira off-flavor [16], phenolic perception [17], a change in hoppy aromas [18], and a detected ether taint [19].

Bottle refermentation has been widely used by Belgian craft brewers for its carbonation effect, giving beer the desired effervescence, and also for the associated oxygen consumption, which limits oxidation and the development of related off-flavors [20–22]. About half a million yeast cells per mL are pitched into the beer before bottling, in the presence of added fermentable sugars. The beer is then kept in a warm room (20–28 °C) from two to four weeks.

According to Saison et al. [23], however, refermentation can be damageable, causing loss of flowery, fruity, and ester notes that are highly appreciated by consumers. Long storage can lead to yeast autolysis with release of esterases (deleterious to fruity esters) and to excretion of amino acids, peptides, and short-chain fatty acids [24–28]. When *Brettanomyces* strains are present in the bottle, production of isovaleric, hexanoic, and octanoic acids is especially promoted [29–31].

The aim of the present paper was to assess how bottle refermentation impacts the flavor properties of Belgian craft beers. As bottle refermentation was already known to significantly improve the release of free-hop thiols from cysteine and glutathione conjugates [22], we decided to investigate only non-dry-hopped commercial samples. First, short-chain fatty acids were investigated in 16 bottle-refermented and two unrefermented Belgian craft beers to determine whether they were present above their sensory threshold. In a few selected samples, more flavors were then analyzed through natural aging at 20 °C in the dark. Esters (isoamyl acetate, ethyl hexanoate, and ethyl octanoate) were quantitated by headspace-GC-FID, and most trace aromas were monitored by GC-olfactometry after XAD-2 aroma extraction. Lastly, some cardboard defects (*trans*-2-nonenal) and a few other changes in aroma were evidenced by overall sensory analysis.

2. Materials and Methods

2.1. Materials

Isoamyl acetate (99%), ethyl hexanoate (99%), ethyl octanoate (99%), 2-pentanol (98%), isovaleric acid (99%), hexanoic acid (≥98%), octanoic acid (≥98%), nonanoic acid (99%), and decanoic acid were purchased from Sigma Aldrich GmbH (Bornem, Belgium); n-hexanol from Acros Organics (Geel, Belgium); ethanol (99.8%) from Merck (Darmstadt, Germany); XAD-2 resin from Supelco Inc. (Bellefonte, United States of America); and sodium chloride, copper sulfate (II), and diethyl ether from VWR International (Leuven, Belgium). Authentic standard flavor compounds for olfactometry were of pure grade (purity >98%) from Sigma-Aldrich. Milli-Q water was used (Millipore, Bedford, MA, USA).

2.2. Beer Samples and Aging Procedure

A total of 18 commercial, top-fermented, late-hopped beers (here listed as A–R for reasons of confidentiality) were kindly supplied by Belgian craft brewers. All were bottle-refermented, except beers A and B. Six representative samples (A–F), same lot as above, were further selected for more in-depth investigations through natural aging (20 °C in the dark). The main characteristics of these beers are depicted in Table 1.

Table 1. Main characteristics of the six selected craft beers.

Beer	Alcohol (% vol)	Real Extract (°P)	pH	Bitterness (BU)	Color (EBC)	Sensorial Characteristics
A	6.5	4.6	4.2	15	12.5	Butter, apple, hop, green
B	12.3	6.6	4.4	20	25	Alcohol, banana, cheese, phenols
C *	7.9	5.1	4.5	21	16.5	Butter, sulfur, hop
D *	8.8	5.8	4.4	14	66	Malt, sulfur, green
E *	8.1	4.2	4.4	29	14.5	Lemon, banana, apple, spicy, phenols
F *	7.5	3.7	4.5	24	15.5	Orange, pineapple, spicy, phenols

*: with bottle refermentation.

2.3. Short-chain Fatty Acid Analysis

First, 100 µL of internal standard (IST—1000 mg/L nonanoic acid) was added to 10 mL of beer in a 20-mL vial flask, which was immediately closed and shaken for 10 s. Then, 300 µL of n-hexanol was added before shaking again for 5 min. Compounds were recovered in assembled n-hexanol fractions after 2 successive centrifugations (14,000 rpm) [32]. Next, 1 µL of extract was analyzed on an Agilent 6890N gas chromatograph equipped with a split injector maintained at 200 °C (split ratio = 73.6). The FID (flame ionization detector) was set at 220 °C. Compounds were injected into a CP-Wax 58 column (Agilent, 60 m × 0.32 mm i.d., 0.5-µm film thickness). The carrier gas was nitrogen, and the pressure was set at 60 kPa. The oven temperature was programmed to rise from 125 to 140 °C at 8 °C/min and then to 180 °C at 15 °C/min. Quantitation was done by determining the relative response of each compound to IST (done by standard addition to beer B). Results are expressed as the average of duplicates.

2.4. Static Headspace Analysis of Esters

Prior to analysis, the beers were stored for 2 h at 4 °C to avoid excessive foaming. The whole procedure was carried out in a cold room (4 °C). Then, 40 µL of internal standard (IST, 2500 mg/L 2-pentanol) and 1.9 g of sodium chloride were added to 5 mL of beer in a 20-mL screw vial flask, which was closed immediately and kept closed until analysis. A total of 500 µL of extract were analyzed on a Thermo Finnigan Trace gas chromatograph, equipped with a splitless injector maintained at 250 °C; the split vent was opened 1 min post-injection. The FID detector was set at 260 °C. Compounds were injected into a VF-Wax MS column (Agilent, 60 m × 0.32 mm i.d., 0.5-µm film thickness). The carrier gas was nitrogen, and the pressure was set at 100 kPa. The oven temperature was programmed to rise from 40 to 140 °C at 8 °C/min and then to 180 °C at 15 °C/min. Quantitation was performed by standard addition (relative response of each compound to IST). Results are expressed as the average of duplicates.

2.5. Flavor XAD-2 Extraction and Gas Chromatography—Olfactometry Analytical Conditions

An extraction procedure based on that of Lermusieau et al. [33], was used to recover aroma compounds from beer. First, 4 g of XAD-2 resin were added to 50 mL of beer in a 250-mL flask. The flask was sealed with a Teflon-lined cap and shaken in the dark for 2 h at 200 rpm. After extraction, the contents were poured into a glass column with a coarse frit and Teflon stopcock, and the liquid was drained off, leaving a small bed of resin, which was further rinsed with 100 mL of distilled water (4 × 25 mL). Aroma compounds were then eluted with 40 mL of diethyl ether (2 × 20 mL). The extract was dried with Na_2SO_4 and concentrated to 0.5 mL in a Kuderna-Danish evaporator at 39 °C. A Chrompack CP9001 gas chromatograph equipped with a splitless injector maintained at 250 °C was used for the olfactometry analyses, and the split vent was opened after 0.5 min. Compounds were separated using a wall-coated open-tubular (WCOT) apolar CP SIL5 CB capillary column (Agilent, 50 m × 0.32 mm, 1.2-µm film thickness) connected to a GC-odor port at 250 °C. The eluent was diluted with a large volume of air (20 mL/min) previously humidified using aqueous copper (II) sulfate solution. The oven temperature was programmed from 36 to 85 °C at 20 °C/min, to 145 °C at 1 °C/min, to 250 °C at 3 °C/min, and then to remain constant at 250 °C for 30 min. A volume of 1 µL of extract was injected. Sniffing was performed by two experienced panelists. Serial dilutions were prepared from the initial XAD-2 extract at a ratio of $1:3^n$ in diethyl ether. The dilution factor (FD) was calculated as 3^n, $n + 1$ being the number of dilutions (factor 3) required for no odor to be perceived (Log_3FD values in Table 2 equal to 0, 1, 2, ..., 10, nd if no odor detected for the undiluted extract). The difference between two Log_3FD becomes significant when above 1.

Table 2. GC-olfactometric analysis of XAD-2 extracts issued from beers A, C, E, and F (fresh and after 6 months of storage at 20 °C). RI: retention index; nd: not detected.

			Log$_3$FD							
			A		C		E		F	
Compound	**RI**	**Odor**	**Fresh**	**Aged**	**Fresh**	**Aged**	**Fresh**	**Aged**	**Fresh**	**Aged**
Ethyl butyrate	778	Red fruits	1	1	4	6	nd	3	3	4
3-Methyl-2-buten-1-thiol	809	Garlic, hoppy	7	8	7	8	10	10	10	10
Isovaleric acid	811	Sweat, rancid	2	2	2	2	4	5	5	5
2-Methylbutanoic acid	828	Sweat	2	2	2	2	4	5	3	6
2-Methyl-3-furanthiol	850	Broth	5	7	5	2	10	10	10	10
Isoamyl acetate	854	Fruity, banana	1	1	2	2	nd	nd	nd	nd
Ethyl hexanoate	979	Fruity, candy	2	2	4	4	3	4	4	4
Furaneol	1037	Cotton candy	4	5	6	7	5	5	4	7
Linalool	1089	Flowery, coriander	7	6	6	7	nd	nd	5	2
trans-2-Nonenal	1127	Cardboard	nd	6	nd	5	3	4	4	5
Citronellol	1216	Fruity, flowery	nd	nd	1	4	3	2	nd	8
4-Vinylguaiacol	1294	Clove	3	4	4	4	5	6	4	6
γ-Nonalactone	1327	Coconut	nd	3	0	4	1	2	3	5
Vanillin	1365	Vanilla	0	nd	6	7	nd	4	1	5
β-Damascenone	1374	Fruity, apricot	2	3	4	5	4	4	3	6
4-Vinylsyringol	1543	Clove	3	3	0	3	2	3	5	5

2.6. Sensory Analyses

A group of 10 panelists (all trained scientists, non-smokers, and regular consumers of craft beers, including three women and seven men aged 23–55 years) scored four aging attributes on a scale of 0–4: cardboard, bread, cooked fruit, and dried fruit. A score of 0 meant the panelist did not detect the aroma, whereas a score of 4 meant the aroma was strongly perceived.

3. Results and Discussion

3.1. Short-Chain Fatty Acids

GC analyses revealed considerable variability of short-chain fatty acid profiles among fresh Belgian craft beers (Figure 1). Most samples contained isovaleric acid (Figure 1a) at a concentration above its threshold (1 mg/L) [34]. In all cases, on the other hand, hexanoic (Figure 1b), octanoic (Figure 1c), and decanoic acids (Figure 1d) were below their sensory thresholds (10, 10, and 5 mg/L, respectively [35]). A good correlation ($R^2 = 0.72$) was found, as expected, between the concentrations of hexanoic acid and octanoic acid (Figure 2).

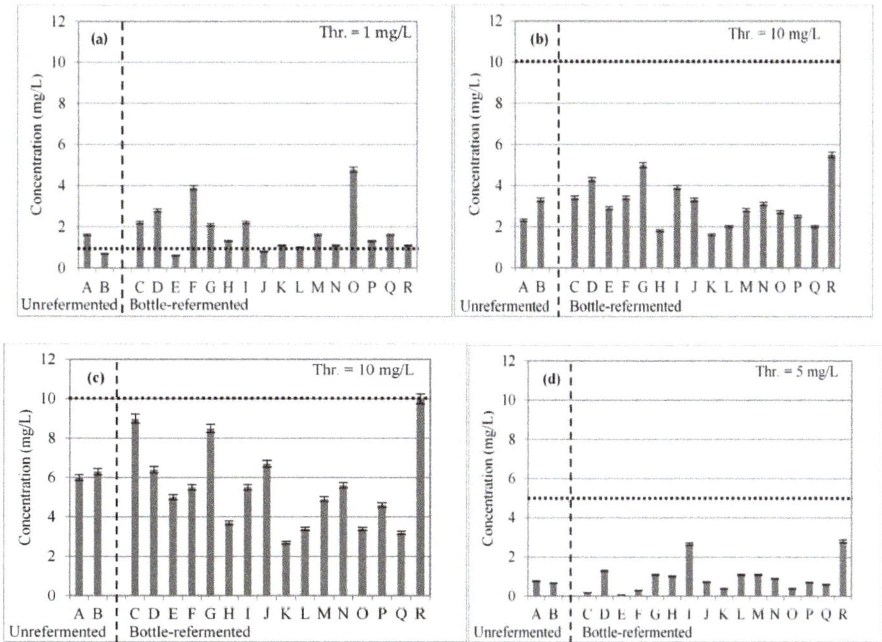

Figure 1. Concentrations of isovaleric (**a**), hexanoic (**b**), octanoic (**c**), and decanoic (**d**) acids (mg/L) in 18 fresh Belgian craft beers. The dotted lines in each graph indicate the compound sensory threshold (Thr.).

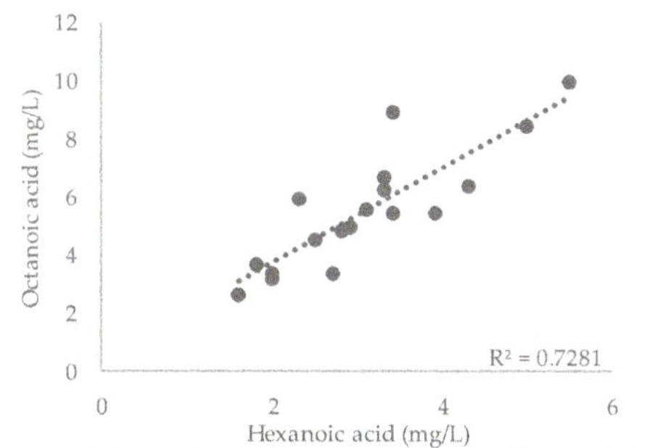

Figure 2. Correlation between the concentrations (mg/L) of octanoic acid and hexanoic acid (in eighteen fresh beers: A–R).

In a representative set of beers (A–F), short-chain fatty acids were determined after 3, 6, and 12 months of natural aging (Figure 3). No increase in isovaleric, hexanoic, octanoic, or decanoic acid concentration was observed over the aging period. Worth mentioning, however, is the significant decrease in octanoic acid in beer C (Figure 3C). This was the only beer tested here in which the concentration of this compound reached 9 mg/L before aging. In conclusion, bottle refermentation does not cause significant release of fatty acids through natural aging and thus does not negatively impact flavor.

Unrefermented

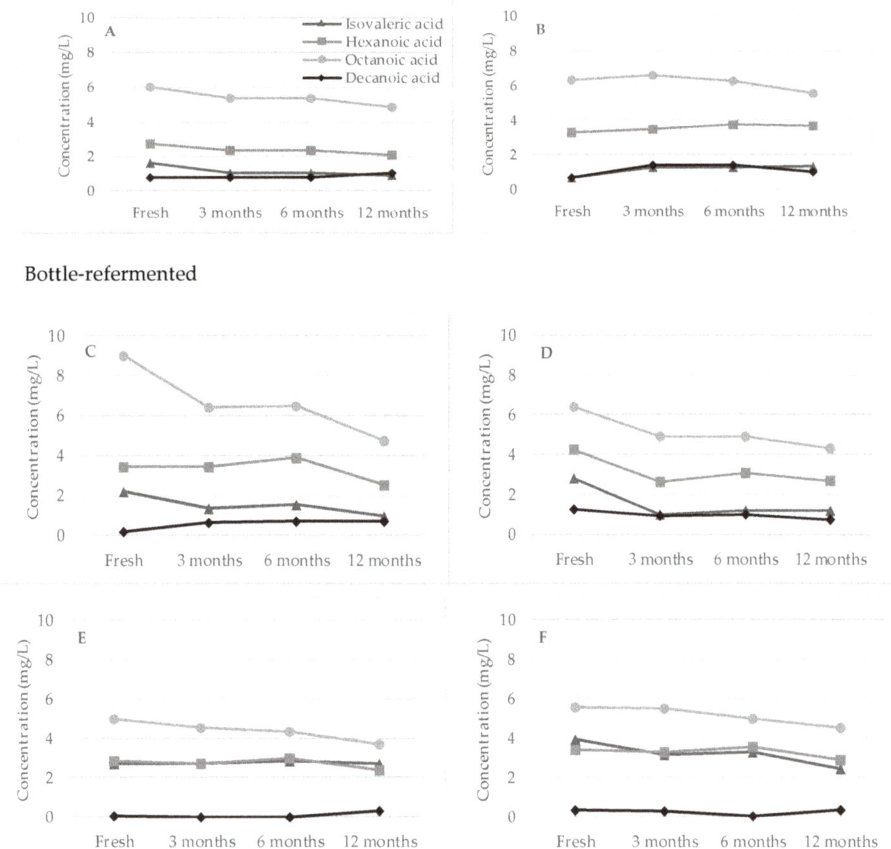

Figure 3. Concentrations (mg/L) of isovaleric acid (▲), hexanoic acid (■), octanoic acid (●), and decanoic acid (♦) in fresh beers (**A–F**) and their evolution during natural aging (3, 6, and 12 months). Variation coefficients under 5%.

3.2. Esters

Esters were determined on the same sampling of craft beers (A–F) in the course of one year of natural aging (Figure 4). For fresh beers, a strong correlation ($R^2 = 0.86$) was again observed between the concentrations of ethyl hexanoate and ethyl octanoate (Figure 5a), although these ester levels did not correlate well with the hexanoic and octanoic acids concentrations (Figure 5b,c). Ethyl hexanoate and ethyl octanoate, were found very close to their sensory thresholds (0.2 and 0.9 mg/L, respectively [36]) in all fresh samples and remained relatively stable during aging, with no significant difference between the unrefermented (A and B) and bottle-refermented (C–F) samples. On the other hand, the fruity banana-like isoamyl acetate was found to be partially degraded throughout the year of storage in all beers except B (characterized by a much higher level of ethanol, Table 1). The similar trend observed in A and C–F confirms that esters can be broken down even without the release of esterases upon yeast autolysis. In all of the beers, the isoamyl acetate concentration remained above the sensory threshold (0.5 mg/L [37]) after one year.

Unrefermented

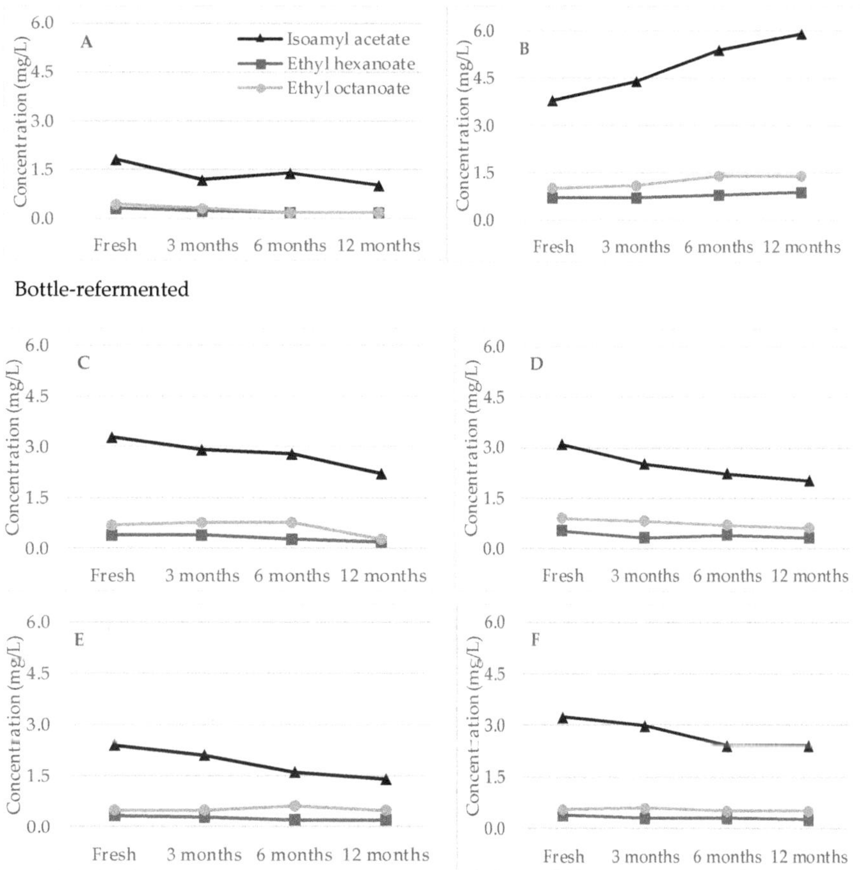

Figure 4. Concentrations (mg/L) of isoamyl acetate (▲), ethyl hexanoate (■), and ethyl octanoate (●) in fresh beers (**A–F**) and their evolution during natural aging (3, 6, and 12 months). Variation coefficients under 5%.

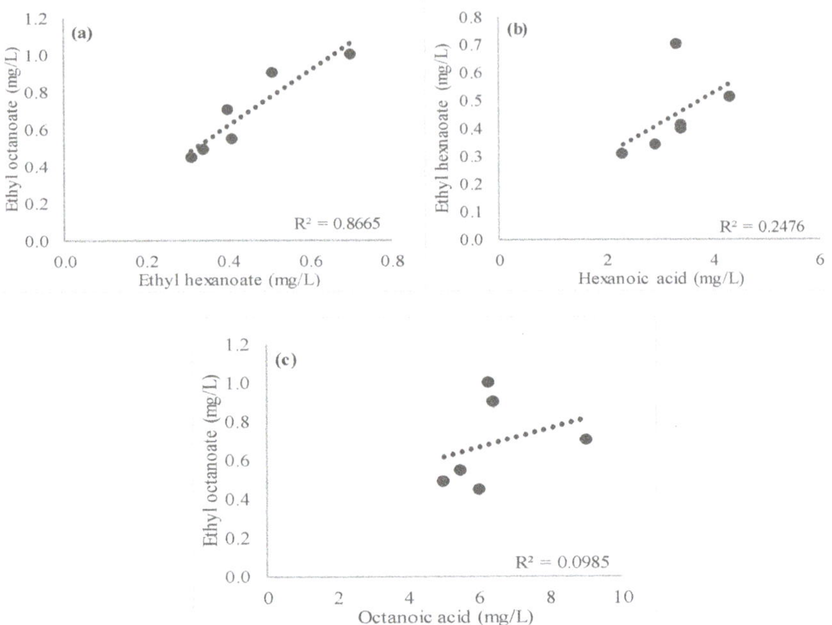

Figure 5. Correlations between the concentrations (mg/L) (**a**) ethyl octanoate and ethyl hexanoate, (**b**) ethyl hexanoate and hexanoic acid, and (**c**) ethyl octanoate and octanoic acid (in six fresh beers: A–F).

3.3. Olfactive Analysis of XAD-2 Extracted Flavors

Samples A, C, E, and F (all blond beers to avoid the complexity of special malt-derived molecules largely investigated elsewhere [5,17]) were subjected to XAD-2 resin extraction followed by GC-olfactometric (GC-O) analysis. Beer odor intensities were determined by the aroma extract dilution analysis (AEDA) [38]. To focus on beer flavor-active compounds, we list in Table 2 only those compounds whose FD was as high as that of ethyl hexanoate, an ester known to be present in the samples at concentrations close to its sensory threshold.

3-Methyl-2-buten-1-thiol (Log_3FD = 7–10, a pleasant hoppy flavor here but also known as skunky at a much higher level [22]), 2-methyl-3-furanthiol (Log_3FD = 2–10, broth), furaneol (Log_3FD = 4–7, cotton candy), and linalool (Log_3FD = 2–7, flowery/coriander) emerged as the most potent odorants in all four beers. The persistent detection of 4-vinylguaiacol (Log_3FD = 3–6) in all samples indicates that POF(+) strains had been used by the brewers.

As already mentioned above, even in bottle-refermented beers (C, E, and F), isoamyl acetate (Log_3FD = 1–2, undetectable in E and F due to the strong previous odor) and ethyl hexanoate (Log_3FD = 2–4) showed good stability through aging, with no significant changes in FD. On the other hand, the red-fruit ethyl butyrate was produced during storage in the bottle-refermented beers (from not detected to Log_3FD = 3 in beer E and from 4 to 6 in beer C).

Although *trans*-2-nonenal showed an increase in all four beers during aging, the highest FD jump was observed for the unrefermented A (from not detected to Log_3FD = 6).

An interesting result was the strong increase, especially in aged beer F, of compounds suspected to be released during storage through glucoside hydrolysis: citronellol, 4-vinylguaiacol, vanillin, and β-damascenone [39,40]. In this case, the selected yeast clearly brought new flavors to F, explaining why some consumers may prefer the six-month-aged beer. The efficiency of the yeast β-1,3-glucanase or β-glucosidase should be taken into account to predict the amounts in which aglycons can be released through aging. Linalool (detected in fresh beers A, C, and F) was the sole hop terpenol found to decrease in beer F (Log_3FD from 5 to 2), suggesting yeast terpenol biotransformations in the bottle [41].

3.4. Sensory Analyses

Beers A–F were investigated by a trained panel while fresh and then after 3 and 6 months of storage. As suspected from the GC-olfactometry results, unrefermented beers A and B developed a relatively intense cardboard flavor (*trans*-2-nonenal), already strongly perceived after 3 months (Figure 6). Despite the increase in FD for furaneol and β-damascenone after 6 months of storage, the attributes bread, dried fruit, and cooked fruit remained absent or relatively weak.

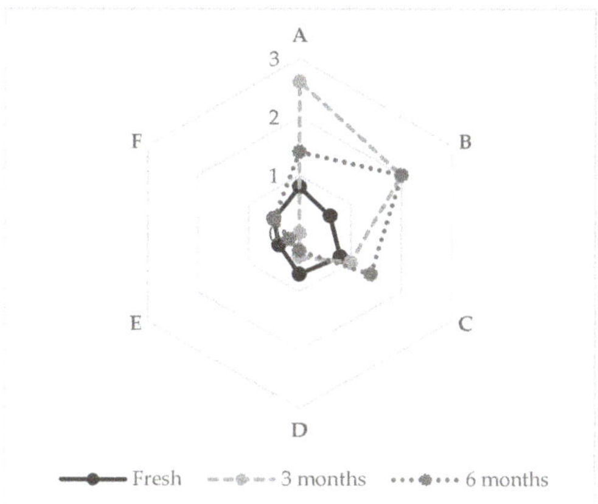

Figure 6. Spider diagram of cardboard flavor intensity in fresh (●), and naturally aged (3 and 6 months) beers (A–F).

4. Conclusions

Bottle refermentation of craft beers can be promoted for its ability both to protect beer against oxidation, (this protection is required to avoid colloidal instability, bitterness decrease, and aroma loss) and to avoid the accumulation of *trans*-2-nonenal through enzymatic reduction to nonenol and nonanol. Moreover, the presence of yeast in the bottle proved not so deleterious as it regards fatty acid excretion and ester hydrolysis during the first year of storage, while leading to the release of interesting terpenols and phenols from aglycons.

Author Contributions: Conceptualization, S.C.; Data curation, C.S.F., E.B., and S.C.; Formal analysis, C.S.F., E.B., and S.C.; Writing—original draft, C.S.F. and S.C.

Funding: This research received no dedicated funding.

Conflicts of Interest: The authors declare no conflicts of interest.

References

1. Pires, E.J.; Teixeira, J.A.; Brányik, T.; Vicente, A.A. Yeast: The soul of beer's aroma—A review of flavour-active esters and higher alcohols produced by the brewing yeast. *Appl. Microbiol. Biotechnol.* **2013**, *98*, 1937–1949. [CrossRef] [PubMed]
2. Saerens, M.G.; Verstrepen, K.; Thevelein, J.; Delvaux, F. Ethyl esters production during brewery fermentation: A review. *Cerevisia* **2008**, *33*, 82–90.
3. Coghe, S.; Benoot, K.; Delvaux, F.; Vanderhaegen, B.; Delvaux, F.R. Ferulic acid release and 4-vinylguaiacol formation during brewing and fermentation: Indications for feruloyl esterase activity in *Saccharomyces cerevisiae*. *J. Agric. Food Chem.* **2004**, *52*, 602–608. [CrossRef] [PubMed]

4. Coghe, S.; Martens, E.; D'Hollander, H.; Dirinck, P.J.; Delvaux, F. Sensory and instrumental flavour analysis of wort brewed with dark specialty malts. *J. Inst. Brew.* **2004**, *110*, 94–103. [CrossRef]

5. Vandecan, S.M.G.; Daems, N.; Schouppe, N.; Saison, D.; Delvaux, F. Formation of flavor, color, and reducing power during the production process of dark specialty malts. *J. Am. Soc. Brew. Chem.* **2011**, *69*, 150–157. [CrossRef]

6. Gros, J.; Peeters, F.; Collin, S. Occurrence of odorant polyfunctional thiols in beers hopped with different cultivars. First evidence of an S-cysteine conjugate in hop (*Humulus lupulus* L.). *J. Agric. Food Chem.* **2012**, *60*, 7805–7816. [CrossRef]

7. Kankolongo Cibaka, M.-L.; Ferreira, C.S.; Decourrière, L.; Lorenzo-Alonso, C.-J.; Bodart, E.; Collin, S. Dry hopping with the dual-purpose varieties Amarillo, Citra, Hallertau Blanc, Mosaic, and Sorachi Ace: minor contribution of hop terpenol glucosides to beer flavors. *J. Am. Soc. Brew. Chem.* **2017**, *75*, 122–129.

8. Kankolongo Cibaka, M.-L.; Decourrière, L.; Lorenzo-Alonso, C.-J.; Bodart, E.; Robiette, R.; Collin, S. 3-Sulfanyl-4-methylpentan-1-ol in dry-hopped beers: first evidence of glutathione S-Conjugates in hop (*Humulus lupulus* L.). *J. Agric. Food Chem.* **2016**, *64*, 8572–8582. [CrossRef]

9. Dalgliesh, C.E. Flavour stability. In Proceedings of the European Brewery Convention: Proceedings of the 16th Congress, Amsterdam, The Netherlands, 1977; pp. 623–659.

10. Noël, S.; Liegeois, C.; Lermusieau, G.; Bodart, E.; Badot, C.; Collin, S. Release of deuterated nonenal during beer aging from labeled precursors synthesized in the boiling kettle. *J. Agric. Food Chem.* **1999**, *47*, 4323–4326. [CrossRef]

11. Lermusieau, G.; Noël, S.; Liegeois, C.; Collin, S. Nonoxidative mechanism for development of *trans*-2-nonenal in beer. *J. Am. Soc. Brew. Chem.* **1999**, *57*, 29–33. [CrossRef]

12. Zufall, C.; Racioppi, G.; Gasparri, M.; Franquiz, J. Flavour stability and ageing characteristics of light stable beers. In Proceedings of the European Brewery Convention Congress, Prague, Czech Republic, 2–5 October 2005; p. 30.

13. Silva Ferreira, C.; Thibault de Chanvalon, E.; Bodart, E.; Collin, S. Why humulinones are key bitter constituents only after dry hopping: comparison with other Belgian styles. *J. Am. Soc. Brew. Chem.* **2018**, *76*, 236–246. [CrossRef]

14. Clapperton, J.F. Ribes flavor in beer. *J. Inst. Brew.* **1976**, *82*, 175–176. [CrossRef]

15. Thi Thu, H.T.; Nizet, S.; Gros, J.; Collin, S. Occurrence of the ribes odorant 3-sulfanyl-3-methylbutyl formate in aged beers. *Flavour Fragr. J.* **2013**, *28*, 147–179.

16. Scholtes, C.; Nizet, S.; Collin, S. How sotolon can impart a Madeira off-flavor to aged beers. *J. Agric. Food Chem.* **2015**, *63*, 2886–2892. [CrossRef] [PubMed]

17. Scholtes, C.; Nizet, S.; Collin, S. Guaiacol and 4-methylphenol as specific markers of torrefied malts. Fate of volatile phenols in special beers through aging. *J. Agric. Food Chem.* **2014**, *62*, 9522–9528. [CrossRef] [PubMed]

18. Thi Thu, H.T.; Kankolongo Cibaka, M.-L.; Collin, S. Polyfunctional thiols in fresh and aged Belgian special beers: fate of hop S-cysteine conjugates. *J. Am. Soc. Brew. Chem.* **2015**, *73*, 61–70.

19. Vanderhaegen, B.; Neven, H.; Daenen, L.; Verstrepen, K.J.; Verachtert, H.; Derdelinckx, G. Furfuryl ethyl ether: important aging flavor and a new marker for the storage conditions of beer. *J. Agric. Food Chem.* **2004**, *52*, 1661–1668. [CrossRef]

20. Derdelinckx, G.; Vanderhasselt, B.; Maudoux, M.; Dufour, J.P. Refermentation in bottles and kegs: A rigorous approach. *Brauwelt Int.* **1992**, *2*, 156–164.

21. Vanderhaegen, B.; Neven, H.; Coghe, S.; Verstrepen, K.J.; Verachtert, H.; Derdelinckx, G. Evolution of chemical and sensory properties during aging of top-fermented beer. *J. Agric. Food Chem.* **2003**, *51*, 6782–6790. [CrossRef]

22. Nizet, S.; Gros, J.; Peeters, F.; Chaumont, S.; Robiette, R.; Collin, S. First evidence of the production of odorant polyfunctional thiols by bottle refermentation. *J. Am. Soc. Brew. Chem.* **2013**, *71*, 15–22. [CrossRef]

23. Saison, D.; De Schutter, D.P.; Vanbeneden, N.; Daenen, L.; Delvaux, F.; Delvaux, F.R. Decrease of aged beer aroma by the reducing activity of brewing yeast. *J. Agric. Food Chem.* **2010**, *58*, 3107–3115. [CrossRef] [PubMed]

24. Ernest, C.-H.; Chen, A.; Jamieson, A.M.; Van Gheluwe, G. The release of fatty acids as a consequence of yeast autolysis. *J. Am. Soc. Brew. Chem.* **1980**, *38*, 13–17.

25. Leroy, M.J.; Charpentier, M.; Duteurtre, B.; Feuillat, M.; Charpentier, C. Yeast autolysis during Champagne aging. *Am. J. Enol. Vitic.* **1990**, *41*, 21–28.

26. Masschelein, C.A. The biochemistry of maturation. *J. Inst. Brew.* **1986**, *92*, 213–219. [CrossRef]

27. Ormrod, I.H.L.; Lalor, E.F.; Sharpe, F.R. The release of yeast proteolytic enzymes into beer. *J. Inst. Brew.* **1991**, *97*, 441–443. [CrossRef]

28. Neven, H.; Delvaux, F.; Derdelinckx, G. Flavor evolution of top fermented beers. *MBAA Tech. Q.* **1997**, *34*, 115–118.

29. Gassenmeier, K.; Schieberle, P. Potent aromatic compounds in the crumb of wheat bread (French-type)-influence of pre-ferments and studies on the formation of key odorants during dough processing. *Z. Lebensm. Unters. Forsch.* **1995**, *201*, 241–248. [CrossRef]

30. Olsen, E. *Brettanomyces*: Occurrence, flavour effects and control. In Proceedings of the 23rd Annual New York Wine Industry Workshop, New York, NY, USA, 23–24 March 1994.

31. Fugelsang, K.C. *Wine Microbiology*; The Chapman & Hall Enology Library: New York, NY, USA, 1997.

32. Alvarez., P.; Malcorps, P.; Sa Almeida, A.; Ferreira, A.; Meyer, A.M.; Dufour, J.P. Analysis of free fatty acids, fusel alcohols, and esters in beer: an alternative to CS2 extraction. *J. Am. Soc. Brew. Chem.* **1994**, *52*, 127–134. [CrossRef]

33. Lermusieau, G.; Bulens, M.; Collin, S. Use of GC–Olfactometry to identify the hop aromatic compounds in beer. *J. Agric. Food Chem.* **2001**, *49*, 3867–3874. [CrossRef]

34. Licker, J.L.; Acree, T.E.; Henick-Kling, T. What is "Brett" (*Brettanomyces*) flavour? A preliminary investigation. In *Chemistry of Wine Flavour*; ACS Symposium Series; American Chemical Society: Washington, DC, USA, 1998; pp. 96–115.

35. Meilgaard, M. Flavor and threshold of beer volatiles. *MBAA Tech. Q.* **1974**, *11*, 87–89.

36. Engan, S. Organoleptic threshold of some alcohols and esters in beer. *J. Inst. Brew.* **1971**, *78*, 33–36. [CrossRef]

37. Saison, D.; De Schutter, D.P.; Uyttenhove, B.; Delvaux, F.; Delvaux, F.R. Contribution of staling sompounds to the aged flavor of lager beer by studying their flavor thresholds. *Food Chem.* **2009**, *114*, 1206–1215. [CrossRef]

38. Ullrich, F.; Grosch, W. Identification of the most intensive volatile flavour compounds formed during autoxidation of linoleic acid. *Z. Lebensm. Unters. Forsch.* **1987**, *184*, 277–282. [CrossRef]

39. Callemien, D.; Dasnoy, S.; Collin, S. Identification of a stale-beer-like odorant in extracts of naturally aged beer. *J. Agric. Food Chem.* **2006**, *54*, 1409–1413. [CrossRef] [PubMed]

40. Chevance, F.; Guyot, C.; Dupont, J.; Collin, S. Investigation of the β-damascenone in fresh and aged commercial beers. *J. Agric. Food Chem.* **2002**, *50*, 3818–3821. [CrossRef] [PubMed]

41. King, A.J.; Dickinson, J.R. Biotransformation of hop aroma terpenoids by ale and lager yeasts. *FEMS Yeast Res.* **2003**, *3*, 53–62. [CrossRef] [PubMed]

MDPI

St. Alban-Anlage 66

4052 Basel

Switzerland

Tel. +41 61 683 77 34

Fax +41 61 302 89 18

www.mdpi.com

Beverages Editorial Office

E-mail: beverages@mdpi.com

www.mdpi.com/journal/beverages